Pearson Edexcel GCSE (9–1)
Mathematics
Foundation tier
Revision Guide + App

Series Consultant: Harry Smith
Author: Harry Smith

A note from the publisher

In order to ensure that this resource offers high-quality support for the associated Pearson qualification, it has been through a review process by the awarding body. This process confirms that this resource fully covers the teaching and learning content of the specification or part of a specification at which it is aimed. It also confirms that it demonstrates an appropriate balance between the development of subject skills, knowledge and understanding, in addition to preparation for assessment.

Endorsement does not cover any guidance on assessment activities or processes (e.g. practice questions or advice on how to answer assessment questions), included in the resource nor does it prescribe any particular approach to the teaching or delivery of a related course.

While the publishers have made every attempt to ensure that advice on the qualification and its assessment is accurate, the official specification and associated assessment guidance materials are the only authoritative source of information and should always be referred to for definitive guidance.

Pearson examiners have not contributed to any sections in this resource relevant to examination papers for which they have responsibility.

Examiners will not use endorsed resources as a source of material for any assessment set by Pearson.

Endorsement of a resource does not mean that the resource is required to achieve this Pearson qualification, nor does it mean that it is the only suitable material available to support the qualification, and any resource lists produced by the awarding body shall include this and other appropriate resources.

> **For the full range of Pearson revision titles across KS2, 11+, KS3, GCSE, Functional Skills, AS/A Level and BTEC visit:**
> www.pearsonschools.co.uk/revise

A small bit of small print

Pearson Edexcel publishes Sample Assessment Material and the Specification on its website. This is the official content and this book should be used in conjunction with it. The worked examples and questions in this book have been written to help you practise the topics in the specification, and to help you prepare for your exams. Remember that the real exam questions may not look like this, and the questions in this book will not appear in your exams.

Your new *Revision Guide* is packed with features to help you stay ahead of the game, and on track for success in your Pearson Edexcel Foundation Maths GCSE.

Examiners' report

Every year Pearson Edexcel produces reports on the most recent exams. These **examiners' reports** are jam-packed with useful advice about which questions students struggled with, where marks were dropped or picked up, and which skills students need to concentrate on. We've taken a deep dive into these reports to bring you the most relevant advice for your upcoming exams. Whenever you see this feature in the *Revision Guide*, you know that you're looking at advice based on **real students** who have sat **real exams**. To get you started, here are our **top five** examiners' tips and tricks.

Five sure-fire ways to pick up marks

1 Write clearly – make sure all your numbers can be read clearly, and don't write so small that the examiner can't read your working.

2 Show all your workings – you can pick up loads of marks for partially correct answers, but only if you have shown your method neatly.

3 Use your calculator – every year students lose marks on calculator papers by using mental or written methods when they could just use their calculator!

4 Don't rush – if there are figures given in the question make sure you copy them carefully into your own working or onto your calculator.

5 Answer the question – always make sure your answer matches what is asked for in the question. Give conclusions in words or sentences, and write numerical answers on the answer line.

Problem solving

Problem solved! You'll need to use problem-solving skills in your exam. Look out for problem-solving tips and strategies wherever you see this icon, and check out the dedicated problem-solving skills pages at the end of each main section.

Target grades

Target grade 5 The exam-style questions in this book have been given target grades. You can use these target grades to help you track your progress. Remember that being able to answer questions at a particular target grade does not guarantee that you will achieve this grade! Your actual grade will be based on the total number of marks you get. So if you are aiming for a top grade you need to be confident with every topic, and you should attempt all the questions in your exams.

Video solutions

 Some of the questions in this *Revision Guide* have worked solution videos. Use your phone, tablet or webcam to scan the QR code to watch the video.

INSIDE TRACK

 Some topics come up year after year. We've picked 25 of the hottest topics. These pages contain key skills and knowledge that you're likely to need in your upcoming exams. If you're pushed for time you might want to revise these first.

 There is some tough material in Foundation GCSE. We've identified 25 of the trickiest topics. Don't worry if you find these pages difficult — they are! You might want to save these topics for days when you have a bit more time to concentrate on them.

Date: ....
Time: ....
Location: ....

Date: ....
Time: ....
Location: ....

Date: ....
Time: ....
Location: ....

You will have to sit **three papers** for Pearson Edexcel Foundation GCSE Maths. Each paper is **1 hour and 30 minutes long** and is worth **80 marks**. You can't use a calculator on Paper 1, but make sure you have one with you for Papers 2 and 3. If you know when and where your exams are taking place, write this information into your Revision Guide here.

The grade boundaries change every year, so you can't know exactly how many marks you need to make your target grade. But you can use this table to get a rough idea.

Target grade	Marks needed
1	15%
2	30%
3	45%
4	60%
5	75%

In it to win it!

Many successful Foundation GCSE students pick up **a few marks** on **lots of questions**, even if they don't get them fully correct. Pick up cheeky marks by:

 showing a bit of working

✓ demonstrating a method

✓ using mathematical language

✓ writing down a formula.

Whatever you do, **have a go!**

iii

CONTENTS

1-to-1 page match with the Mathematics Revision Workbook ISBN 9781447987925

- ii-iii Get the inside track!

NUMBER
- 1 Place value
- 2 Negative numbers
- 3 Rounding numbers
- 4 Adding and subtracting
- 5 Multiplying and dividing
- 6 Decimals and place value
- 7 Operations on decimals
- 8 Squares, cubes and roots
- 9 Indices
- 10 Estimation
- 11 Factors, multiples and primes
- 12 HCF and LCM
- 13 Fractions
- 14 Operations on fractions
- 15 Mixed numbers
- 16 Calculator and number skills
- 17 Standard form 1
- 18 Standard form 2
- 19 Counting strategies
- 20 Problem-solving practice 1
- 21 Problem-solving practice 2

ALGEBRA
- 22 Collecting like terms
- 23 Simplifying expressions
- 24 Algebraic indices
- 25 Substitution
- 26 Formulae
- 27 Writing formulae
- 28 Expanding brackets
- 29 Factorising
- 30 Linear equations 1
- 31 Linear equations 2
- 32 Inequalities
- 33 Solving inequalities
- 34 Sequences 1
- 35 Sequences 2
- 36 Coordinates
- 37 Gradients of lines
- 38 Straight-line graphs 1
- 39 Straight-line graphs 2
- 40 Real-life graphs
- 41 Distance–time graphs
- 42 Rates of change
- 43 Expanding double brackets
- 44 Quadratic graphs
- 45 Using quadratic graphs
- 46 Factorising quadratics
- 47 Quadratic equations
- 48 Cubic and reciprocal graphs
- 49 Simultaneous equations
- 50 Rearranging formulae
- 51 Using algebra
- 52 Identities and proof
- 53 Problem-solving practice 1
- 54 Problem-solving practice 2

RATIO & PROPORTION
- 55 Percentages
- 56 Fractions, decimals and percentages
- 57 Percentage change 1
- 58 Percentage change 2
- 59 Ratio 1
- 60 Ratio 2
- 61 Metric units
- 62 Reverse percentages
- 63 Growth and decay
- 64 Speed
- 65 Density
- 66 Other compound measures
- 67 Proportion
- 68 Proportion and graphs
- 69 Problem-solving practice 1
- 70 Problem-solving practice 2

GEOMETRY & MEASURES
- 71 Symmetry
- 72 Quadrilaterals
- 73 Angles 1
- 74 Angles 2
- 75 Solving angle problems
- 76 Angles in polygons
- 77 Time and timetables
- 78 Reading scales
- 79 Perimeter and area
- 80 Area formulae
- 81 Solving area problems
- 82 3-D shapes
- 83 Volumes of cuboids
- 84 Prisms
- 85 Units of area and volume
- 86 Translations
- 87 Reflections
- 88 Rotations
- 89 Enlargements
- 90 Pythagoras' theorem
- 91 Line segments
- 92 Trigonometry 1
- 93 Trigonometry 2
- 94 Exact trigonometry values
- 95 Measuring and drawing angles
- 96 Measuring lines
- 97 Plans and elevations
- 98 Scale drawings and maps
- 99 Constructions 1
- 100 Constructions 2
- 101 Loci
- 102 Bearings
- 103 Circles
- 104 Area of a circle
- 105 Sectors of circles
- 106 Cylinders
- 107 Volumes of 3-D shapes
- 108 Surface area
- 109 Similarity and congruence
- 110 Similar shapes
- 111 Congruent triangles
- 112 Vectors
- 113 Problem-solving practice 1
- 114 Problem-solving practice 2

PROBABILITY & STATISTICS
- 115 Two-way tables
- 116 Pictograms
- 117 Bar charts
- 118 Pie charts
- 119 Scatter graphs
- 120 Averages and range
- 121 Averages from tables 1
- 122 Averages from tables 2
- 123 Line graphs
- 124 Stem-and-leaf diagrams
- 125 Sampling
- 126 Comparing data
- 127 Probability 1
- 128 Probability 2
- 129 Relative frequency
- 130 Frequency and outcomes
- 131 Venn diagrams
- 132 Set notation
- 133 Independent events
- 134 Problem-solving practice 1
- 135 Problem-solving practice 2
- 136 ANSWERS

KEY

 = Hot Topic

 = Tricky Topic

Had a look ☐ Nearly there ☐ Nailed it! ☐

NUMBER

Place value

The value of each digit in a number depends on its position. Digits that are further to the left are worth more. You can use a place value diagram to help you read and write numbers.

Worked example
Target grade 1

Write these amounts in order, smallest first.
(1 mark)

£2908   £2950   £5011   £925   £10 430
£925
£925   £10 430
£925   £5011   £10 430
£925   £2908   £2950   £5011   £10 430

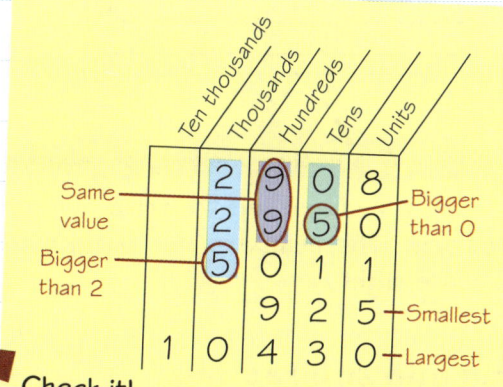

Check it!
Have you included all the amounts in your final answer? ✓

Working with money

 Do all your calculations in the same units, either £ or p.
✓ Write either £ or p in your answer, but not both.
✓ 100p = £1
✓ Amounts in pounds need 2 decimal places. Write 280p as £2.80.

Worked example
Target grade 1

A music website sells songs and albums.
Songs cost 79p each.
Albums cost £6.99 each.
Aaron has a £25 gift card.
He buys 2 albums and spends the rest on songs.
How many songs can he afford?   **(2 marks)**
2 × 6.99 = 13.98
25 − 13.98 = 11.02
11.02 ÷ 0.79 = 13.949…
Aaron can buy 13 songs.

Examiners' report

You can get credit for attempting the correct calculation, so make sure you show all your working. Write neatly, and copy any numbers from the question carefully.
Make sure you read the question carefully – your final answer needs to be a **number of songs**, not an amount of money.
Choose whether you want to work in pounds or pence, and remember to round down because Aaron can only buy a **whole number** of songs.

Real students have struggled with questions like this in recent exams – **be prepared!**

Now try this
Target grade 1

1 Write these amounts in order, smallest first:
   £1974   £974   £1749   £1497   £1947
   (1 mark)

Profit is money you make, and **loss** is money you lose. You need to **add** the individual profits, then **subtract** the individual losses.

2 Mark buys and sells used cars. The table shows information about five cars that Mark bought and sold.

Car	A	B	C	D	E
bought	£2200	£7800	£4200	£……	£6500
sold	£2900	£9000	£……	£11 200	£5750
profit or loss	£700 profit	£…… ……	£1500 loss	£500 profit	£…… ……

(a) Complete the table. **(2 marks)**
(b) Work out Mark's total profit or loss for these five cars. **(2 marks)**

NUMBER Had a look ☐ Nearly there ☐ Nailed it! ☐

Negative numbers

Numbers less than than 0 are called **negative** numbers.

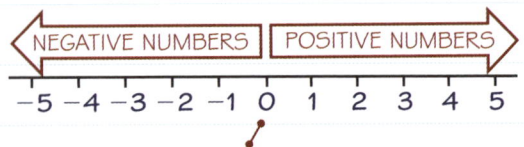

0 is neither positive nor negative.

You can use a number line to write numbers in order of size. The numbers get bigger as you move to the right.

Number lines

You can use number lines to help when adding and subtracting.

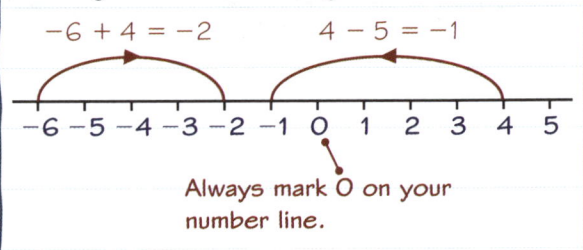

Always mark 0 on your number line.

Adding and subtracting

To add or subtract a negative number, change the double signs first.

$+ - \rightarrow -$
$12 + -3 = 12 - 3 = 9$

$- - \rightarrow +$
$5 - -9 = 5 + 9 = 14$

Golden rule

When you **add** a negative number the answer is **lower**.
When you **subtract** a negative number the answer is **higher**.

Multiplying and dividing

When multiplying and dividing, use these rules to decide whether the answer will be positive or negative.

1 If numbers have the **same** sign then the answer is **positive**.
$-3 \times -7 = 21$

2 If numbers have **different** signs then the answer is **negative**.
$+80 \div -10 = -8$

Problem solved!
One strategy is to try a few pairs and add up the totals. You could also work out what each total should be by adding together all 6 cards then dividing by 3:
$-4 + -3 + 0 + 2 + -6 + -1 = -12$
$-12 \div 3 = -4$

Check it!
Make sure you have used each card exactly once. ✓

Worked example Target grade 1

Here are six number cards.

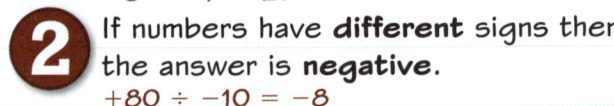

Sort the cards into three pairs so that each pair has the same total. **(2 marks)**

−4 and 0   Total = −4
−3 and −1 Total = −4
−6 and 2   Total = −4

Now try this

1 (a) Work out $-2 + -4$
  (b) Work out $-1 - -6$
   (2 marks)

2 (a) Work out -4×-12
  (b) Work out $30 \div -5$
   (2 marks)

Worked solution video

3 Here are six number cards.

Sort the cards into three pairs so that each pair has the same total. **(2 marks)**

Target grade 1

Try a few pairs and add up the totals.

Had a look ☐   Nearly there ☐   Nailed it! ☐   **NUMBER**

Rounding numbers

To **round** a number, you look at the next digit to the right on a place value diagram.
5 or more → round up, less than 5 → round down

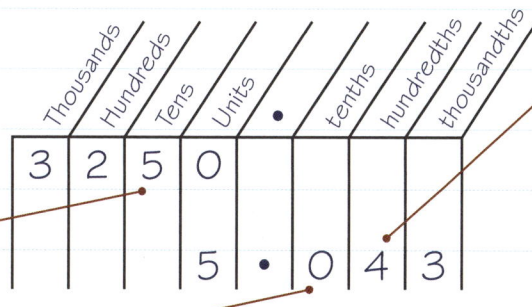

To round to the nearest 100, you look at the digit in the tens column. It is a 5, so round up. 3250 rounded to the nearest 100 is 3300.

To round to 1 decimal place (1 d.p.), you look at the digit in the second decimal place. It is a 4, so round down. 5.043 rounded to 1 d.p. is 5.0 You **need** to write the 0 to show that you have rounded to 1 d.p.

To round to the nearest whole number look at the digit in the tenths column. It is a 0, so round down. 5.043 rounded to the nearest whole number is 5.

Significant figures

You always start counting **significant figures** from the first non-zero on the left.

27.05 rounded to 1 s.f. is 30 — The first non-zero digit is 2. The next digit is 7, so round up to give an answer of 30.
27.05 rounded to 2 s.f. is 27 — Look for the two digits furthest to the left, which are 2 and 7. The next digit is 0, so round down to give an answer of 27.
27.05 rounded to 3 s.f. is 27.1 — The first three significant figures are 2, 7 and 0. The next digit is 5 so round up to 27.1.

Numbers less than 1

When rounding numbers less than 1 to a given number of significant figures, remember **not** to count zero digits that are on the left.

0.0085 rounded to 1 s.f. is 0.009 — Look for the digit which is furthest to the left and which is **not** a zero. This digit is 8. The next digit is 5 so round up to give an answer of 0.009.

Worked example

Target grade 3

Round 6.9083 correct to
(a) 1 significant figure
(b) 2 significant figures
(c) 3 significant figures. **(3 marks)**

(a) 7 (1 s.f.)
(b) 6.9 (2 s.f.)
(c) 6.91 (3 s.f.)

The width of the cooker is given to the nearest whole unit. So it might be **inaccurate** by up to half a unit in either direction. Show Supraj is wrong by giving an example of a width larger than 76 cm that would round to 76 cm.
There is more about rounding errors on page 32.

Now try this

 **1** Round the number 3756
  (a) to the nearest 100 **(1 mark)**
  (b) to the nearest 10 **(1 mark)**

 **2** Write these numbers correct to 1 significant figure:
  (a) 48 **(1 mark)**
  (b) 3025 **(1 mark)**
  (c) 0.00939 **(1 mark)**
  (d) 6.5 **(1 mark)**

Worked solution video

 **3** Round 0.179554 to 3 significant figures. **(1 mark)**

 **4** Supraj is installing a kitchen. He orders a cooker that is 76 cm wide, to the nearest cm. He says: 'If I leave a gap exactly 76 cm wide the cooker will definitely fit perfectly.' Explain why Supraj is wrong. **(1 mark)**

NUMBER

Had a look ☐ Nearly there ☐ Nailed it! ☐

Adding and subtracting

You need to be able to add and subtract numbers without a calculator.

Mental methods

Try these methods for adding and subtracting quickly in your head.

213 + 79

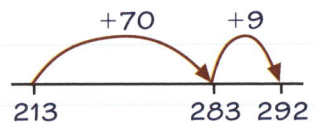

213   283 292

Add the tens first then the units.
213 + 79 = 292

152 − 63

63 70   100   152

Count on in steps from 63 up to 152. Add up the steps to work out the difference between 152 and 63.
7 + 30 + 52 = 89
152 − 63 = 89

Worked example — Target grade 1

Work out 285 + 56 + 1091   **(2 marks)**

```
   285
    56
+ 1091
------
  1432
   2 1
```

1. **Always** add the units column first: 5 + 6 + 1 = 12. Write down the 2 and carry the **1** over to the tens column.
2. Add the tens column: 8 + 5 + 9 + **1** = 23. Make sure you include any numbers you carried over. Write down the 3 and carry the **2** over to the hundreds column.
3. Add the hundreds column: 2 + 0 + **2** = 4. Write down 4.
4. There is only one digit in the thousands column. Write this in your answer.

Worked example — Target grade 1

Work out 418 − 62   **(2 marks)**

```
  ³⁄₄̸18
 − 62
-----
   356
```

1. **Always** subtract the units column first: 8 − 2 = 6.
   Remember it is (top number) − (bottom number).
2. Look at the tens column. 1 is smaller than 6 so you have to exchange 1 hundred for 10 tens. Change 4 hundreds into 3 hundreds and 10 tens.
3. Now you can subtract the tens column: 11 − 6 = 5.
4. There is nothing to subtract in the hundreds column so write 3 in your answer.

Now try this — Target grade 1

1. Work out
   - (a) 503 + 1126 + 85   **(2 marks)**
   - (b) 745 + 283   **(2 marks)**

2. (a) Work out 627 − 251   **(2 marks)**
   (b) How much is 831 − 659?   **(2 marks)**

3. Joe buys a magazine costing £4.45 and two birthday cards costing £1.99 each. He pays with a £10 note. How much change will he receive?   **(3 marks)**

Scan this QR code to watch a video of this question being solved.

You can work in pence so you don't have to use decimal numbers. Work out 445 + 199 + 199, then subtract the result from 1000. Remember to give units with your answer.

Had a look ☐ **Nearly there** ☐ **Nailed it!** ☐   **NUMBER**

Multiplying and dividing

You need to be able to multiply and divide numbers without a calculator.
For a reminder about multiplying and dividing by 10, 100 and 1000 have a look at page 61.

Mental methods

Try these methods for multiplying and dividing quickly in your head.

37×8
$30 \times 8 = 240$
$7 \times 8 = 56$
$37 \times 8 = 296$

Split 37 into 30 and 7. Then multiply each by 8. Add the answers to get the total.

$54 \div 6$
$6 \times \square = 54$
The answer is 9.

Try to find a multiplication fact using 6 with 54 as the answer.

Worked example   Target grade 1

Work out
(a) 49×3 **(2 marks)**   (b) 36×24 **(3 marks)**

```
    49
 ×   3
 ----
   147
    2
```

```
    36
 ×  24
 ----
   144
    2
   720
    1
 ----
   864
```

Always multiply from right to left.
1. $9 \times 3 = 27$. Write down 7 and carry over **2** (2 tens).
2. $4 \times 3 = 12$. Add on the carry-over. $12 + 2 = 14$. Write down 14.

Examiners' report

Remember that this is a **non-calculator** question. You need to show all your working.
1. Work out 36×4. (Answer = 144)
2. Work out 36×20. Write down **0** and then work out 36×2. (Answer = 72**0**)
3. Add the answers. (144 + 720 = 864)

Multiplying and dividing are **much easier** if you are confident with your **times tables** up to 10×10.

Real students have struggled with questions like this in recent exams – **be prepared!**

Worked example   Target grade 1

Work out $288 \div 9$ **(2 marks)**

```
      32
   _____
 9 )288
   -27
   ---
    18
   -18
   ---
     0
```

You can use long division.
1. Does 9 divide into 2? No.
2. Does 9 divide into 28? Yes. $9 \times 3 = 27$ so 9 divides into 28 three times with remainder 1.
3. Does 9 divide into 18? Yes. $9 \times 2 = 18$ so 9 divides into 18 two times with no remainder.

Using short division the calculation would look like this:

```
      3 2
   _____
 9 )28¹8
```

Now try this

 Target grade 1

1. (a) Work out 72×100 **(1 mark)**
   (b) Work out 256×9 **(1 mark)**
   (c) Work out 29×78 **(2 marks)**

Worked solution video

 Target grade 1

2. (a) Work out $468 \div 3$ **(2 marks)**
   (b) Work out $1032 \div 8$ **(2 marks)**

Target grade 2

3. There are 18 chocolate coins in a bag.
   Paula buys 6 of these bags.
   Paula has 7 grandchildren. She wants to give each of her grandchildren 15 coins.
   Has she bought enough coins? **(3 marks)**

You need to answer 'yes' or 'no' to the question, **and** show your working to **justify** your answer.

5

| NUMBER | Had a look ☐ | Nearly there ☐ | Nailed it! ☐ |

Decimals and place value

You can use a place value diagram to help you understand and compare decimal numbers. Remember that decimal numbers with more digits are not necessarily bigger. Try writing extra 0s so that all the numbers have the same number of decimal places.

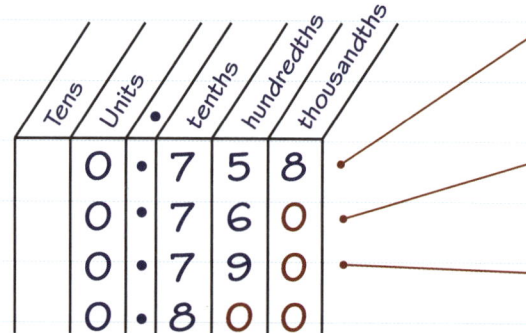

The value of the 5 in this number is 5 hundredths.

0.76 is the same as 0.760
0.76 is bigger than 0.758
6 hundredths is bigger than 5 hundredths.

0.79 is smaller than 0.8 because the digit in the tenths place is smaller.

Worked example — Target grade 1

Write these numbers in order, smallest first:
0.43   0.425   0.4   0.48   0.459   (2 marks)

~~0.43~~ ~~0.425~~ ~~0.4~~ ~~0.48~~ ~~0.459~~
0.4   0.425
0.4   0.425   0.43
0.4   0.425   0.43   0.459
0.4   0.425   0.43   0.459   0.48

All these numbers have the same tenths digit. You need to look at the hundredths digit first. 0.4 is the same as 0.40 so this is the smallest number.
0.425 has the next smallest hundredths digit so this is the next number.

Check it!
Cross out each number to make sure you include them all in your final answer. ✓

Worked example — Target grade 2

Using the information that
58 × 71 = 4118
write down the value of
(a) 58 × 0.71   (1 mark)
41.18
(b) 5800 × 7.1   (1 mark)
41180

(a) 71 has been divided by 100 and 58 hasn't been changed. So the answer needs to be divided by 100.   4118 ÷ 100 = 41.18

(b) 58 has been multiplied by 100 and 71 has been divided by 10.
—[×100]—[÷10]→ is the same as —[×10]→
The answer needs to be multiplied by 10:
4118 × 10 = 41180
To revise multiplying and dividing by 10, 100 and 1000 see page 61.

Now try this

1 Write down the place value of the 8 in these numbers:
   (a) 2.84 **(1 mark)**   (b) 0.3086 **(1 mark)**

2 Write these numbers in order of size:
   0.517   0.508   0.58   0.571   0.51   **(2 marks)**

3 Using the information that
   672 × 13 = 8736
   write down the value of
   (a) 0.672 × 13   **(1 mark)**
   (b) 8736 ÷ 67.2   **(1 mark)**

You could start with the smallest number or the largest.

Had a look ☐ Nearly there ☐ Nailed it! ☐   **NUMBER**

Operations on decimals

1. Adding and subtracting

To add or subtract decimal numbers:
1. Line up digits with the same place value.
2. Line up the decimal points.
3. Write a decimal point in your answer.

See page 4 for a reminder about adding and subtracting.

> Write in 0s so that both numbers have the same number of decimal places.

Worked example — Target grade 1

(a) $0.75 + 1.6$ (1 mark)

```
  0.75
+ 1.60
------
  2.35
   1
```

(b) $3.5 - 0.21$ (1 mark)

```
  3.⁴5̸¹0
- 0.2 1
-------
  3.2 9
```

2. Multiplying

To multiply decimal numbers:
1. Ignore the decimal points and just multiply the numbers.
2. Count the number of decimal places in the calculation.
3. Put this number of decimal places in the answer.

> (a) You can use estimation to check that the decimal point is in the correct place.
> $8.69 \times 12 \approx 9 \times 12 = 108 \quad 108 \approx 104$ ✓
>
> (b) 8.5×0.04 has 3 decimal places in total
> So $8.5 \times 0.04 = 0\overset{\frown}{.}3\overset{\frown}{4}\overset{\frown}{0} = 0.34$
>
> Write a 0 before the decimal point and simplify your answer.

Worked example

(a) 8.69×12 — Target grade 1 (2 marks)

```
    869
 ×   12
-------
  1 738
    1 1
+ 8690
-------
 10428
   1 1
```

$8.69 \times 12 = 104.28$

(b) 8.5×0.04 — Target grade 2 (2 marks)

```
   85
 ×  4
-----
  340
   2
```

$8.5 \times 0.04 = 0.34$

3. Dividing

To divide by a decimal number:
1. Multiply both numbers by 10, 100 or 1000 to make the second number a whole number.
2. Divide by the whole number.

> Multiply 40.6 and 1.4 by 10.
> If you multiply both numbers in a division by the same amount, the answer stays the same.

Worked example — Target grade 2

(a) $55.8 \div 3$ (1 mark)

```
    1 8.6
   _____
 3)5²5.¹8
```

(b) $40.6 \div 1.4$ (2 marks)

$40.6 \div 1.4 = 406 \div 14$

```
     29
   ____
14)406
   -28
   ---
    126
   -126
   ----
      0
```

$40.6 \div 1.4 = 29$

Now try this

1 (a) Work out $12.5 + 7.93$ (2 marks)
   (b) Work out $8.14 + 3 + 0.772$ (2 marks)

2 (a) Work out $16.5 - 9.72$ (2 marks)
   (b) Work out 5.76×34 (2 marks)

3 (a) A kitchen stool costs £39.90.
       Work out the cost of 6 kitchen stools. (2 marks)
   (b) Tom buys 12 pencils for £11.28.
       Work out the cost of 1 pencil. (2 marks)

Worked solution video

> Make sure you line up the decimal points.

7

NUMBER

Had a look ☐   Nearly there ☐   Nailed it! ☐

Squares, cubes and roots

Squares and square roots

When a whole number is multiplied by itself the answer is a **square number**. You can write square numbers using index notation.

Multiplication	Index notation	Square number
2 × 2	2^2	4
5 × 5	5^2	25
9 × 9	9^2	81
13 × 13	13^2	169

Square numbers are the areas of squares with whole number side lengths.
There is more about area on page 80.

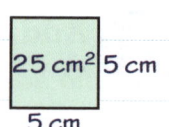

Square roots are the opposite of squares. You use the symbol $\sqrt{}$ to represent a square root.

$\sqrt{4} = 2$ $\sqrt{25} = 5$ $\sqrt{81} = 9$

You need to be able to **remember** the square numbers up to 15 × 15 and the corresponding square roots.

Cubes and cube roots

When a whole number is multiplied by itself then multiplied by itself again, the answer is a **cube number**. You can write cube numbers using index notation.

Multiplication	Index notation	Cube number
2 × 2 × 2	2^3	8
3 × 3 × 3	3^3	27
5 × 5 × 5	5^3	125
10 × 10 × 10	10^3	1000

Cube numbers are the volumes of cubes with whole number side lengths.
There is more about volume on page 83.

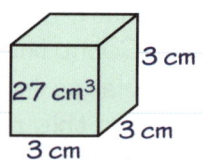

Cube roots are the opposite of cubes. You use the symbol $\sqrt[3]{}$ to represent a cube root.

$\sqrt[3]{8} = 2$ $\sqrt[3]{27} = 3$ $\sqrt[3]{1000} = 10$

You need to be able to **remember** the cubes of 2, 3, 4, 5 and 10, and the corresponding cube roots.

Worked example

Pavel says
'A square number cannot be a prime number.'
Is Pavel correct? Explain your answer. (2 marks)

Yes because a prime number has exactly two factors: 1 and itself. The square number 1 has only one factor (1) so it is not prime. Every other square number has at least three factors (1, itself and its square root) and so is not prime.

Problem solved!

If a question says 'explain' you must answer the question and give a **reason**. Start your answer with: 'Yes because...' or 'No because...'. If you're not sure you could **try some numbers** to see what is going on:

Factors of 9 are 1, 3, 9, so 9 is **not prime**.
Factors of 49 are 1, 7, 49, so 49 is **not prime**.

Revise factors and prime numbers on page 11.

Now try this

1 Work out
   (a) 6^2 **(1 mark)**
   (b) $\sqrt[3]{125}$ **(1 mark)**
   (c) 4^3 **(1 mark)**
   (d) $\sqrt{121}$ **(1 mark)**

 You should know these without using your calculator.

2 Amy says
'The product of two square numbers is always an even number.'
Give an example to show that Amy is wrong. **(1 mark)**

Had a look ☐ Nearly there ☐ Nailed it! ☐   **NUMBER**

Indices

Indices include square roots, cube roots and powers. You can use the **index laws** to simplify powers.

1 $a^m \times a^n = a^{m+n}$
$4^3 \times 4^7 = 4^{3+7} = 4^{10}$

2 $\dfrac{a^m}{a^n} = a^{m-n}$
$12^8 \div 12^3 = 12^{8-3} = 12^5$

3 $(a^m)^n = a^{mn}$
$(7^3)^5 = 7^{3 \times 5} = 7^{15}$

4 Negative powers
$a^{-n} = \dfrac{1}{a^n}$
$5^{-2} = \dfrac{1}{5^2} = \dfrac{1}{25}$

Be careful!
A **negative** power can still have a **positive** answer.

5 Reciprocals
$a^{-1} = \dfrac{1}{a}$

This means that a^{-1} is the **reciprocal** of a. You can find the reciprocal of a fraction by turning it upside down.
$\left(\dfrac{5}{9}\right)^{-1} = \dfrac{9}{5}$

6 Powers of fractions
$\left(\dfrac{a}{b}\right)^n = \dfrac{a^n}{b^n}$
$\left(\dfrac{3}{10}\right)^2 = \dfrac{3^2}{10^2} = \dfrac{9}{100}$

7 Combining rules
You can apply the rules one at a time.
$\left(\dfrac{a}{b}\right)^{-n} = \left(\dfrac{b}{a}\right)^n = \dfrac{b^n}{a^n}$
$\left(\dfrac{2}{3}\right)^{-3} = \left(\dfrac{3}{2}\right)^3 = \dfrac{3^3}{2^3} = \dfrac{27}{8}$

Powers of 0 and 1
Anything raised to the power 0 is equal to 1.
$6^0 = 1 \quad 1^0 = 1 \quad 7223^0 = 1 \quad (-5)^0 = 1$
Anything raised to the power 1 is equal to itself.
$8^1 = 8 \quad 499^1 = 499 \quad (-3)^1 = -3$

Worked example *Target grade 4*

(a) Write $6 \times 6 \times 6 \times 6 \times 6$ as a single power of 6. **(1 mark)**
$6 \times 6 \times 6 \times 6 \times 6 = 6^5$

(b) Simplify $\dfrac{3^7 \times 3}{3^4}$ fully, leaving your answer in index form. **(2 marks)**
$\dfrac{3^7 \times 3}{3^4} = \dfrac{3^8}{3^4} = 3^4$

3 is the same as 3^1. For part (b), use the rule $a^m \times a^n = a^{m+n}$ to simplify the numerator, then use $\dfrac{a^m}{a^n} = a^{m-n}$ to simplify the fraction. Remember to write down both steps of your working and give your answer as a power.

Now try this

Target grade 4

1 Simplify, leaving your answers as a power of 7.
(a) $7^2 \times 7^8$ **(1 mark)**
(b) $7^9 \div 7^2$ **(1 mark)**
(c) $\dfrac{7^4 \times 7^5}{7^6}$ **(2 marks)**
(d) $(7^5)^3$ **(1 mark)**

2 (a) Write $\dfrac{5 \times 5^8}{5^4}$ as a single power of 5. **(2 marks)**

(b) Work out $\dfrac{4 \times 4^{10}}{4^5 \times 4^3}$ **(2 marks)**

Target grade 5

3 Write down the value of
(a) 4^0 **(1 mark)**
(b) 4^{-1} **(1 mark)**
(c) 4^{-2} **(1 mark)**
(d) $\left(\dfrac{2}{5}\right)^2$ **(1 mark)**
(e) $\left(\dfrac{3}{4}\right)^{-3}$ **(1 mark)**

Worked solution video

NUMBER

Had a look ☐ Nearly there ☐ Nailed it! ☐

Estimation

You can estimate the answer to a calculation by rounding each number to **1 significant figure**, and then doing the calculation. You can use this method to check your answers, or to estimate calculations on your **non-calculator paper**. Here are two examples:

1) $4.32 \times 18.09 \approx 4 \times 20 = 80$
The answer is approximately equal to 80.

2) $327^2 \approx 300^2 = 3^2 \times 100^2 = 90\,000$
The answer is approximately equal to 90 000.

$\approx$ means 'is approximately equal to'

Decimal division trick

You might have to divide by a decimal on your non-calculator paper. If you multiply both numbers in a division by the same amount the answer stays the same.

$$\frac{1400}{0.05} = \frac{140\,000}{5} = \frac{280\,000}{10} = 28\,000$$

(×100, ×2 arrows)

Worked example — Target grade 4

Work out an estimate for

(a) $\dfrac{4.31 \times 278}{0.487}$ **(2 marks)**

$\dfrac{4.31 \times 278}{0.487} \approx \dfrac{4 \times 300}{0.5} = \dfrac{1200}{0.5} = 2400$

(b) 37.4^3 **(2 marks)**

$37.4^3 \approx 40^3 = 4^3 \times 10^3 = 64 \times 1000 = 64\,000$

> Round all the numbers to **1 significant figure**. Then **write out** the calculation with the rounded values before calculating your estimate.

> You can use the laws of indices to work out 40^3 without a calculator.
> $(ab)^n = a^n \times b^n$
> so $40^3 = (4 \times 10)^3 = 4^3 \times 10^3$

Problem solved! On your non-calculator paper you can use $\pi = 3.142$ then round to 1 s.f. to make your estimate.

Worked example — Target grade 5

A spherical ball-bearing has a radius of 2.35 cm.

Surface area of sphere = $4\pi r^2$

(a) Work out an estimate for its surface area in square centimetres. **(2 marks)**

$4\pi r^2 = 4 \times 3.142 \times 2.35^2$
$\approx 4 \times 3 \times 2^2 = 48 \text{ cm}^2$

(b) Is your answer to part (a) an overestimate or an underestimate? Give a reason for your answer. **(1 mark)**

$3 < 3.142$ and $2 < 2.35$
so the answer is an underestimate.

Examiners' report

You have rounded **both values down** so your answer will be an underestimate. The question says 'give a reason' so show working and write a conclusion in **words**.

Real students have struggled with questions like this in recent exams – **be prepared!**

Now try this

1 (Target grade 4) Showing your rounding, work out an estimate for

$\dfrac{82 \times 285}{64 \times 35}$ **(2 marks)**

Worked solution video

2 (Target grade 5) A scientist models a raindrop as a sphere with radius 3.2 mm.
Volume of a sphere = $\frac{4}{3}\pi r^3$

(a) Work out an estimate for the volume of the raindrop. **(2 marks)**

(b) Is your answer to part (a) an overestimate or an underestimate? Give a reason for your answer. **(1 mark)**

Had a look ☐   Nearly there ☐   Nailed it! ☐   **NUMBER**

Factors, multiples and primes

Factors and multiples

The **factors** of a number are any whole numbers that divide into it exactly.

1 and the number itself are both factors of any number.
The factors of 12 are 1, 2, 3, 4, 6 and 12.

Factors come in pairs. Each pair is a multiplication with the number as its answer.
The factor pairs of 12 are 1 × 12, 2 × 6 and 3 × 4.

A common factor is a number that is a factor of two or more numbers.
2 is a common factor of 6 and 12.

The **multiples** of a number are all the numbers in its times table.
The multiples of 7 are 7, 14, 21, 28, 35, …

A common multiple is a number that is a multiple of two or more numbers.
12 is a common multiple of 6 and 4.

Primes

A **prime number** has exactly two factors. It can only be divided by 1 and by itself.
The first ten prime numbers are
2, 3, 5, 7, 11, 13, 17, 19, 23, 29.
1 is not a prime number. It has only 1 factor.

Factor trees

You can use a factor tree to find prime factors.
1. Choose a factor pair of the number.
2. Circle the prime factors as you go along.
3. Continue until every branch ends with a prime number.
4. At the end write down **all** the circled numbers, putting in multiplication signs.

Worked example   Target grade 1

Here is a list of numbers:
16   8   3   17   6   20   12
From this list write down
(a) a prime number   **(1 mark)**
17
(b) a multiple of 5   **(1 mark)**
20
(c) two factors of 24 which have a sum of 15.   **(2 marks)**
12 and 3

(a) 3 is also a prime number.
(c) 3, 6, 8 and 12 are all factors of 24. Only 12 and 3 have a sum of 15.

Worked example   Target grade 4

Write 90 as the product of prime factors. Give your answer in index form.   **(3 marks)**

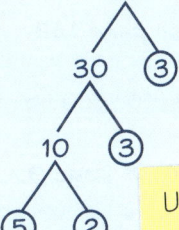

$90 = 5 \times 2 \times 3 \times 3$
$= 5 \times 2 \times 3^2$

Use a factor tree to find all the prime factors. 3 appears twice in the factor tree, so you have to write 3^2.

Check it!
$5 \times 2 \times 3^2 = 10 \times 9 = 90$ ✓

Now try this

1 Here is a list of numbers:
2   8   15   18   21   24   37   44
From the list, write down
(a) the number that is a multiple of 7   **(1 mark)**
(b) the number that is a factor of 45   **(1 mark)**
(c) the number that is a multiple of 6 and a factor of 48   **(1 mark)**
(d) two prime numbers.   **(2 marks)**

2 Write 280 as the product of its prime factors.
Give your answer in index form.   **(3 marks)**

Make sure you can recognise all the prime numbers below 50.

NUMBER

Had a look ☐ Nearly there ☐ Nailed it! ☐

HCF and LCM

The **highest common factor (HCF)** of two numbers is the highest number that is a factor of both numbers.

The **lowest common multiple (LCM)** of two numbers is the lowest number that is a multiple of both numbers.

Worked example *Target grade 4*

(a) Express 108 as a product of powers of its prime factors. **(3 marks)**

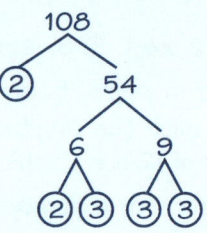

$108 = 2 \times 2 \times 3 \times 3 \times 3$
$= 2^2 \times 3^3$

(b) $240 = 2^4 \times 3 \times 5$
Find, as a product of powers of its prime factors
(i) the highest common factor (HCF) of 108 and 240 **(1 mark)**

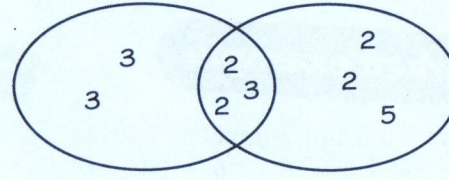

Factors of 108   Factors of 240

$HCF = 2 \times 2 \times 3 = 2^2 \times 3$

(ii) the lowest common multiple (LCM) of 108 and 240. **(1 mark)**

$LCM = 2 \times 2 \times 2 \times 2 \times 3 \times 3 \times 3 \times 5$
$= 2^4 \times 3^3 \times 5$

Examiners' report

If you have to write a number as a **product** of prime factors, make sure you use × signs in your final answer. Don't use +, and don't just write a list of prime factors.

Real students have struggled with questions like this in recent exams – **be prepared!**

You can find the HCF and LCM by writing the products of prime factors in a **Venn diagram**. Use the powers to tell you how many times each prime factor occurs. Put the **common factors** in the intersection of the two ovals.

- HCF = product of all the prime factors in the intersection
- LCM = product of all the prime factors in the Venn diagram

There is more on Venn diagrams on page 132.

You can use **index notation** (powers) to simplify your answers.
For a reminder about this look at page 9.

Worked example *Target grade 4*

n is a number.
10 is the highest common factor of 20 and *n*.
Work out two different possible values for *n*.
(2 marks)

10 and 30

You can use prime factors to answer this question.
$20 = 2 \times 2 \times 5$
$10 = 2 \times 5$
Any number which contains the prime factors 2 and 5 exactly once will work.

Now try this *Target grade 4*

1. (a) Write 132 and 110 as products of powers of their prime factors. **(4 marks)**
   (b) Find the HCF of 132 and 110. **(1 mark)**
   (c) Find the LCM of 132 and 110. **(1 mark)**

Worked solution video

2. Given that $X = 3^2 \times 5 \times 7^4$ and $Y = 2 \times 3^3 \times 5^2$ write down, as a product of powers of its prime factors,
   (a) the HCF of X and Y **(1 mark)**
   (b) the LCM of X and Y. **(1 mark)**

Expand the index notation, then draw a Venn diagram showing the prime factors of each number.

12

Had a look ☐   Nearly there ☐   Nailed it! ☐   **NUMBER**

Fractions

You need to be able to work confidently with fractions, with or without a calculator.

1 Dividing objects

You can use fractions to divide an object into parts.

$\frac{2}{3}$ of this rectangle is shaded.

The top number is called the **numerator**.

The bottom number is called the **denominator**.

2 Equivalent fractions

Different fractions can describe the same amount.

$\frac{1}{2} = \frac{2}{4}$

$\frac{1}{2}$ and $\frac{2}{4}$ are called equivalent fractions.

You can find equivalent fractions by multiplying or dividing the numerator and denominator by the same number.

3 Cancelling fractions

To **cancel** or **reduce** a fraction you divide the top and bottom by the same number.

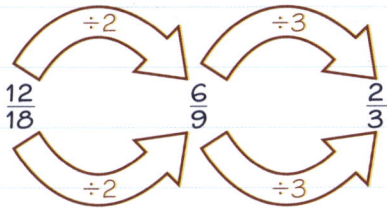

When you can't cancel the fraction any further it is in its **simplest form**.

4 Finding a fraction of an amount

Divide the amount by the denominator
↓
Multiply by the numerator

Work out $\frac{3}{10}$ of 200 kg:

200 kg ÷ 10 = 20 kg

20 kg × 3 = 60 kg

To see how to convert between fractions and decimals look at page 56.

Worked example [Target grade 1]

(a) Write $\frac{20}{80}$ as a fraction in its simplest form.

$\frac{20}{80} = \frac{2}{8} = \frac{1}{4}$ **(1 mark)**

(b) Work out $\frac{2}{5}$ of £240.

240 ÷ 5 = 48

48 × 2 = 96

$\frac{2}{5}$ of £240 is £96   **(2 marks)**

Examiners' report

It is **much easier** to find equivalent fractions if you are confident with your times tables. Write down all the steps when you are simplifying.

Real students have struggled with questions like this in recent exams – **be prepared!**

$\frac{2}{5}$ is less than 1 so the answer should be less than £240. ✓

Plan your strategy before you start. Work out how much money he makes in total, compare this with £8 and then write a conclusion.

Now try this [Target grade 2]

1 David buys 80 chocolates for £20.
He sells $\frac{3}{4}$ of the chocolates for 40p each.
He then sells the remaining chocolates for 20p each.
Work out the total profit that David makes.   **(4 marks)**

2 Sandeep buys some pencils to sell at the school fete.
He buys 40 pencils for £8.
He sells a quarter of the pencils for 25p each and sells half of the pencils for 30p each.
The remaining pencils are broken and are thrown away.
Does Sandeep make a profit or loss?
You must show your working.   **(4 marks)**

NUMBER

Had a look ☐   Nearly there ☐   Nailed it! ☐

Operations on fractions

Make sure you can add, subtract, multiply and divide fractions **without** a calculator.

1 Adding or subtracting

- Write both fractions as equivalent fractions with the same denominator
- ↓
- Add or subtract the numerators
- ↓
- Do not change the denominator

Worked example — Target grade 2

Work out
(a) $\frac{1}{5} + \frac{3}{10}$ (2 marks)   (b) $\frac{8}{9} - \frac{1}{6}$ (2 marks)

$= \frac{2}{10} + \frac{3}{10}$ $= \frac{16}{18} - \frac{3}{18}$

$= \frac{5}{10} = \frac{1}{2}$ $= \frac{13}{18}$

18 is the lowest common multiple (LCM) of 9 and 6. This is the easiest common denominator to use. **For a reminder about LCMs see page 12.**

2 Multiplying

- Write any whole numbers on their own as fractions with denominator 1
- ↓
- Multiply the numerators and multiply the denominators

Worked example — Target grade 2

Work out
(a) $\frac{2}{3} \times \frac{7}{10}$ (2 marks)   (b) $3 \times \frac{2}{11}$ (2 marks)

$= \frac{2 \times 7}{3 \times 10}$ $= \frac{3}{1} \times \frac{2}{11}$

$= \frac{14}{30} = \frac{7}{15}$ $= \frac{6}{11}$

3 Dividing

- Write any whole numbers on their own as fractions with denominator 1
- ↓
- Turn the second fraction 'upside down'
- ↓
- Change ÷ to ×
- ↓
- Multiply the numerators and multiply the denominators

Worked example — Target grade 2

Work out
(a) $\frac{2}{5} \div \frac{3}{4}$ (2 marks)   (b) $6 \div \frac{2}{3}$ (2 marks)

$= \frac{2}{5} \times \frac{4}{3}$ $= \frac{6}{1} \div \frac{2}{3}$

$= \frac{8}{15}$ $= \frac{6}{1} \times \frac{3}{2}$

Change $\frac{3}{4}$ into $\frac{4}{3}$ and change ÷ to ×.

$= \frac{18}{2}$

$= 9$

Watch out!

1. You do not have to cancel your final answer unless the question asks you to 'give your answer in its simplest form'.
2. You can compare and order fractions by using equivalent fractions with the same denominator.

See page 13 for a reminder about equivalent fractions and simplest form.

Now try this — Target grade 2

Worked solution video

1 Work out
(a) $\frac{5}{9} + \frac{1}{3}$ (2 marks)
(b) $\frac{3}{10} \div \frac{8}{15}$ (2 marks)

2 Work out
(a) $\frac{7}{8} - \frac{1}{2}$ (2 marks)   (b) $\frac{8}{11} - \frac{3}{5}$ (2 marks)

3 Work out
(a) $\frac{5}{9} \times \frac{3}{10}$ (2 marks)   (b) $\frac{8}{15} \div \frac{4}{7}$ (3 marks)

Had a look ☐   Nearly there ☐   Nailed it! ☐   **NUMBER**

Mixed numbers

Mixed numbers have a whole number part and a fraction part.

$3\frac{1}{4}$ This mixed number is the same as $3 + \frac{1}{4}$

Improper fractions have a numerator larger than their denominator.

$\frac{5}{2}$, $\frac{21}{5}$ and $\frac{4}{3}$ are all improper fractions.

Converting between mixed numbers and improper fractions

To convert a mixed number into an improper fraction you...

Multiply this... $3\frac{1}{4} = \frac{3 \times 4 + 1}{4} = \frac{13}{4}$

...by this... $3 \times 4 = 12$ ...add it to this. $12 + 1 = 13$ Keep the same denominator.

To convert an improper fraction into a mixed number you...

Divide this... $\frac{23}{5} = 23 \div 5 = 4\frac{3}{5}$ Write the remainder as the numerator.

...by this.   Keep the same denominator.

Golden rule
You need to write mixed numbers as improper fractions before you do any calculations.

Worked example   Target grade 4

Work out $4\frac{1}{2} \times 3$ **(2 marks)**

$= \frac{9}{2} \times \frac{3}{1}$

$= \frac{27}{2}$

$= 13\frac{1}{2}$

$4\frac{1}{2} = \frac{4 \times 2 + 1}{2} = \frac{9}{2}$

Worked example   Target grade 4

Work out $3\frac{2}{5} - 1\frac{1}{2}$ **(3 marks)**

$= \frac{17}{5} - \frac{3}{2}$

$= \frac{34}{10} - \frac{15}{10} = \frac{19}{10}$

$= 1\frac{9}{10}$

Worked example   Target grade 4

Kumar has $2\frac{1}{2}$ pizzas. He wants to share them equally between 3 people. What fraction of a pizza does each person receive?   **(2 marks)**

$2\frac{1}{2} \div 3 = \frac{5}{2} \div \frac{3}{1}$

$= \frac{5}{2} \times \frac{1}{3}$

$= \frac{5}{6}$

Each person receives $\frac{5}{6}$ of a pizza.

Examiners' report

Working with mixed numbers in word problems is tricky. Here are some helpful tips:
- Convert mixed numbers to improper fractions before calculating.
- If you are sharing then divide.
- Use common sense to check your answers. In this question, Kumar has less than 3 whole pizzas, so each person will get less than 1 whole pizza. You can use this fact to check your answer.

Real students have struggled with questions like this in recent exams – **be prepared!**

Now try this   Target grade 4

1. Work out
   (a) $3\frac{1}{3} \times 1\frac{3}{4}$ **(3 marks)**
   (b) $4\frac{1}{5} \div 1\frac{7}{8}$ **(3 marks)**

Write both numbers as improper fractions before you do the calculation. Remember you can't use a calculator for these questions.

2. A cup holds $\frac{1}{3}$ litre of water.
   How many **full** cups can be filled from a $2\frac{1}{2}$ litre jug of water?   **(3 marks)**

NUMBER Had a look ☐ Nearly there ☐ Nailed it! ☐

Calculator and number skills

These calculator keys are really useful:

- x^2 — Square a number.
- x^3 — Cube a number.
- x^{-1} — Find the reciprocal of a number.
- Ans — Use your previous answer in a calculation.
- (−) — Enter a negative number.
- $\sqrt{\square}$ — Find the square root of a number.
- $\sqrt[3]{\square}$ — Find the cube root of a number. You might need to press the shift key first.
- S⇔D — Change the answer from a fraction or surd to a decimal. Not all calculators have this key.

Order of operations

You need to use the correct **priority of operations when doing a calculation**.

Brackets
Indices
Division
Multiplication
Addition
Subtraction

$(10 - 7) + 4 \times 3^2$
$= 3 + 4 \times 3^2$
$= 3 + 4 \times 9$
$= 3 + 36$
$= 39$

Reciprocals

The reciprocal of a number is 1 divided by that number. You can find it by writing the number as a fraction then turning it upside down.

$7 = \frac{7}{1} \to \frac{1}{7}$ The reciprocal of 7 is $\frac{1}{7}$

$\frac{3}{4} \to \frac{4}{3}$ The reciprocal of $\frac{3}{4}$ is $\frac{4}{3}$

You can use the x^{-1} key on your calculator to find reciprocals.

Worked example — Target grade 4

(a) Work out the value of $\frac{\sqrt{8.3}}{12.5 - 7.3}$

Give your answer as a decimal. Write down all the figures on your calculator display. **(2 marks)**

$\frac{\sqrt{8.3}}{12.5 - 7.3} = \frac{2.88097}{5.2} = 0.554033088$

(b) Find the reciprocal of 12.5. Give your answer as a decimal. **(1 mark)**

$1 \div 12.5 = 0.08$

Read the question carefully. You have to give the answer as a **decimal**, so you might need to use the S⇔D button on your calculator.

Examiners' report

If you have to work out a calculation like this in your exam, you should work out the numerator (top) and denominator (bottom) **separately**, and **write them both down**. Then divide to work out the final answer. Read the question carefully. You have to write down **all the figures** from your calculator display.

Check it!
Check your answer by working out the whole calculation in one go using the  key.

Real students have struggled with questions like this in recent exams – **be prepared!**

Now try this

Try doing these **without a calculator**. Make sure you use the correct priority of operations.

Target grade 1

1 Work out
   (a) $10 - 3 \times 2$ **(1 mark)**
   (b) $84 \div (6 + 6)$ **(1 mark)**
   (c) $8 + 3^2$ **(1 mark)**
   (d) $(9 - 4)^3$ **(1 mark)**

Target grade 2

2 Work out
   (a) $\sqrt{244 + 717}$ **(1 mark)**
   (b) $\sqrt[3]{3.1 - 0.356}$ **(1 mark)**

Use a calculator and give your answers as decimal numbers. You might need to use the S⇔D key to convert a fraction answer to a decimal.

Worked solution video

Had a look ☐   Nearly there ☐   Nailed it! ☐   NUMBER

Standard form 1

Numbers in standard form are written as the product of two parts.

$$7.3 \times 10^{-6}$$

This part is a number greater than or equal to 1 and less than 10

This part is an integer power of 10

You can use standard form to write very large or very small numbers.

$$920\,000 = 9.2 \times 10^5$$

Numbers greater than 10 have a positive power of 10

$$0.007\,03 = 7.03 \times 10^{-3}$$

Numbers less than 1 have a negative power of 10

Counting decimal places

You can count decimal places to convert between numbers in standard form and ordinary numbers.

3 jumps

$$7\,900 = 7.9 \times 10^3$$

$7900 > 10$ so the power is positive

4 jumps

$$0.000\,35 = 3.5 \times 10^{-4}$$

$0.000\,35 < 1$ so the power is negative

Be careful!
Don't just count zeros to work out the power.

Worked example Target grade 5

(a) Write 1 630 000 in standard form. (1 mark)
1.63×10^6

(b) Write 4.2×10^{-3} as an ordinary number. (1 mark)
0.0042

Count the number of decimal places you need to move to get a number between 1 and 10. 1 630 000 is bigger than 10 so the power will be positive.

Using a calculator

You can enter numbers in standard form using the  key.
To enter 3.7×10^{-6} press

If you are using a calculator with numbers in standard form it is a good idea to put brackets around each number.

Your calculator might give the answer as a normal number. You need to convert it into standard form.

Worked example Target grade 5

A and B are standard form numbers.
$A = 1.9 \times 10^4$ $B = 4.2 \times 10^5$
Calculate, giving your answers in standard form

(a) $A + B$ (1 mark)
$(1.9 \times 10^4) + (4.2 \times 10^5) = 439\,000$
$= 4.39 \times 10^5$

(b) $B \div 6$ (1 mark)
$(4.2 \times 10^5) \div 6 = 70\,000 = 7 \times 10^4$

Now try this Target grade 5

1 (a) Write 0.000 28 in standard form. (1 mark)
   (b) Write 3.91×10^5 as an ordinary number. (1 mark)

2 p and q are standard form numbers.
   $p = 3.5 \times 10^6$ $q = 7.9 \times 10^5$
   Calculate, giving your answers in standard form
   (a) $p - q$ (1 mark)
   (b) $5p$ (1 mark)

Standard form 2

This page shows **non-calculator** methods for doing calculations with standard form.

1 Multiplying numbers in standard form

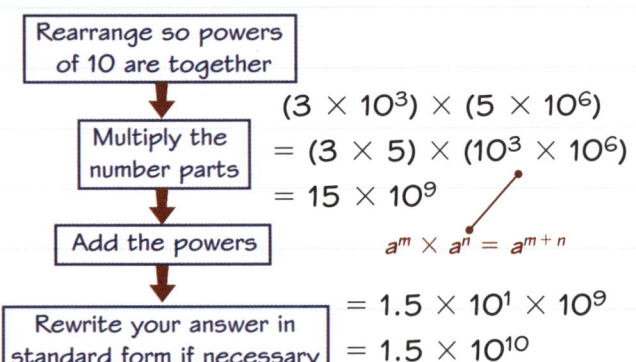

2 Dividing numbers in standard form

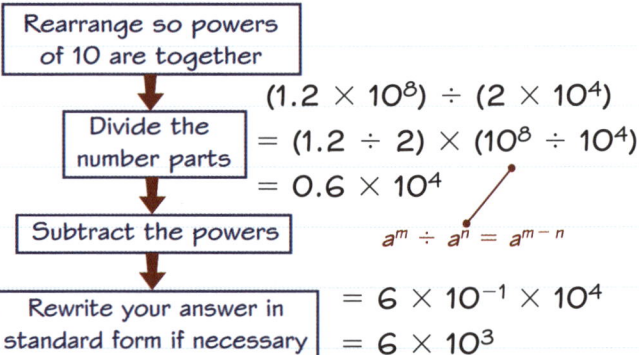

3 Adding and subtracting numbers in standard form

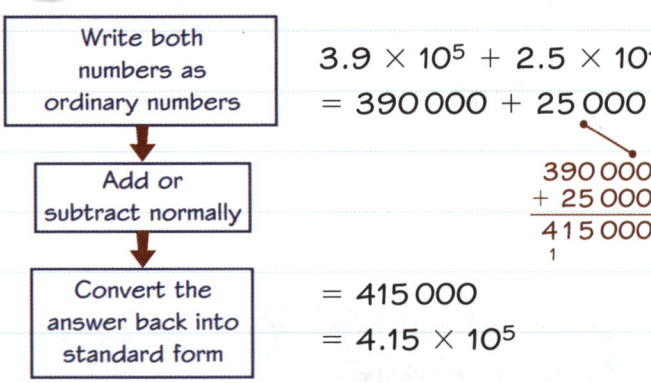

Worked example (Target grade 5)

A and B are standard form numbers.
$A = 8.2 \times 10^6$ $B = 2 \times 10^4$
Calculate $A \times B$ giving your answer in standard form. **(2 marks)**

$(8.2 \times 10^6) \times (2 \times 10^4)$
$= (8.2 \times 2) \times (10^6 \times 10^4)$
$= 16.4 \times 10^{10}$
$= 1.64 \times 10^1 \times 10^{10}$
$= 1.64 \times 10^{11}$

Problem solved!

If you're not sure which operation to use, try using simpler numbers:

An egg weighs 80 g and a box of eggs weighs 960 g. Calculate the number of eggs in the box.

You would divide the total weight by the weight of each egg. So for this worked example, divide the mass of the gold bar by the mass of one atom.

Worked example (Target grade 5)

A single atom of gold has a mass of 3×10^{-22} g.
A gold bar has a mass of 1.2 kg.
Calculate the number of atoms of gold in the bar. **(3 marks)**

$1.2 \text{ kg} = 1200 \text{ g} = 1.2 \times 10^3 \text{ g}$

$\dfrac{1.2 \times 10^3}{3 \times 10^{-22}} = \dfrac{1.2}{3} \times \dfrac{10^3}{10^{-22}}$

$= 0.4 \times 10^{25}$
$= 4 \times 10^{-1} \times 10^{25}$
$= 4 \times 10^{24}$

Now try this (Target grade 5)

1 (a) Work out
   $(3 \times 10^7) \times (9 \times 10^4)$
   Write your answer in standard form. **(2 marks)**

   (b) Work out
   $(6.4 \times 10^{-9}) \div (2 \times 10^{-3})$
   Write your answer in standard form. **(2 marks)**

Had a look ☐   Nearly there ☐   Nailed it! ☐   **NUMBER**

Counting strategies

You might need to make a list of possible combinations. You need to find a **systematic** way of doing this to make sure you have found every possible combination. You could use the three number cards on the right to make **six** different three-digit numbers. Here is one way of writing them out systematically.

[4] [5] [6]

1 4 5 6
2 4 6 5

Start by writing out the numbers that begin with 4.

3 5 4 6
4 5 6 4

Then write the numbers beginning with 5.

5 6 4 5
6 6 5 4

Finally, write the numbers that begin with 6.

Worked example *Target grade 3*

James is making badges to sell. He can choose three different colours:
Red (R)   Green (G)   Blue (B)
He can also choose three shapes:
Circle (C)   Star (S)   Oval (O)
How many possible combinations of badge colour and shape can James make? **(2 marks)**

RC   GC   BC
RS   GS   BS
RO   GO   BO
There are 9 possible combinations.

You need to use a systematic method to make sure you have written down every possible combination. Start by writing down all the possible red badges. Then write down all the possible green badges and then all the possible blue badges.

Make sure you **answer the question**. You need to show a systematic strategy **and** write the number of combinations.

Problem solved! You can save time in your exam by numbering a list of items like this. You can still clearly **show your strategy** using the numbers, and you don't have to write as much.
Read the question carefully. Each team must play the other teams only once, so be careful not to count each match more than once.

You might need to combine counting with probability.
Have a look at page 129 to revise this skill.

Worked example *Target grade 4*

A school hockey league contains five teams:
Benfield ①   Gosforth ②   Kenton ③
St Cuthbert's ④   Sacred Heart ⑤
Each team must play each of the other teams once. How many matches will be played in total?
(2 marks)

1 vs 2, 1 vs 3, 1 vs 4, 1 vs 5
2 vs 3, 2 vs 4, 2 vs 5
3 vs 4, 3 vs 5
4 vs 5   There will be 10 matches in total.

Now try this *Target grade 3*

1 Seth has three tickets to go to the cinema. He chooses two friends to go with him. He chooses one friend from Ali, Ben and Carl, and one friend from Yvonne and Zara.
List all the possible combinations of friends that Seth could choose.
(2 marks)

2 A local darts tournament has six clubs competing for a place in the national finals: Compton, Penn, Goldthorn, Lanesfield, Tettenhall, Wightwick.
Over a four-week period each club must play all the other clubs once each.
How many darts matches will be played in total during the tournament? **(2 marks)**

NUMBER

Had a look ☐ Nearly there ☐ Nailed it! ☐

Problem-solving practice 1

About half of the questions in your Foundation GCSE exam will require you to **problem-solve**, **reason**, **interpret** or **communicate** mathematically. If you come across a tricky or unfamiliar question in your exam you can try some of these strategies:

- ✓ Sketch a diagram to see what is going on.
- ✓ Try the problem with smaller or easier numbers.
- ✓ Plan your strategy before you start.
- ✓ Write down any formulae you might be able to use.
- ✓ Use x or n to represent an unknown value.

AO2 **AO3**

Now try this

1. Wilfred has these coins.

Nisha has these coins.

Wilfred gives Nisha one coin. Wilfred now has twice as much money as Nisha. What value is the coin Wilfred gives to Nisha?

(4 marks)

Place value page 1 *Target grade 1*

In money questions you should decide whether you are going to work in pounds or pence. Don't mix up your units.

Remember that if Wilfred gives a coin to Nisha his amount of money is **reduced** by that value and hers is **increased** by the same amount.

TOP TIP

With a question like this it's easy to check your answer. Write down the new amounts of money Wilfred and Nisha have and check that Wilfred's total is twice Nisha's.

---

2. Liam is planning a trip for a group of 20 children and 5 adults.
   They can go to either the theatre or the zoo.
   If they go to the theatre, they will go by train.
   If they go to the zoo, they will go by coach.
   Liam has information about the costs.

Theatre ticket prices
Stalls: £22
Circle: £15

Return train fares
Adults: £11.50
Child: £5.75

Zoo admission
Adult: £18
Child: £12

Coach hire
20 Seats: £190
30 Seats: £240
40 Seats: £300

   What is the lowest possible total cost of the trip? You must show all your working.

   (5 marks)

Calculator and number skills page 16 *Target grade 2*

Work out the total cost of each trip. Remember to choose the cheapest ticket price for the theatre tickets, and write down all your working. You have to say which trip is the cheapest to complete your answer.

TOP TIP

Plan how you will lay out your answer. You need to show **what** you are working out at each stage, so write short headings to go with your workings. You could use these headings:

Cost of theatre trip
Cost of zoo trip
Conclusion

Had a look ☐   Nearly there ☐   Nailed it! ☐   NUMBER

Problem-solving practice 2

Now try this

3 The diagram shows three identical shapes. $\frac{3}{4}$ of shape A is shaded and $\frac{3}{5}$ of shape C is shaded.

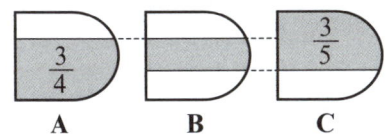

What fraction of shape B is shaded? **(3 marks)**

Fractions page 13

If the whole of each shape is 1, then you can start by working out the fractions that represent the white areas on shapes A and C. You can then subtract these from 1 to find the shaded area on shape B.

TOP TIP

You have to **show your strategy**. This means you need to show all your working so it is clear how you have tackled the problem.

---

4 Susan has 2 dogs.
Each dog is fed $\frac{3}{8}$ kg of dog food each day.
Susan buys dog food in bags.
Each bag weighs 14 kg.
For how many days can Susan feed the 2 dogs from 1 bag of dog food?
You must show **all** your working. **(3 marks)**

Fractions page 13

There are lots of steps in this question so make sure you keep track of your working.

TOP TIP

If your working is very untidy or hard to follow, then re-write it clearly and cross out your original working.

---

5 The diagram shows two types of plastic building block.

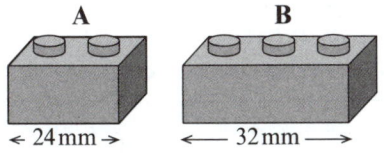

Block A is 24 mm long.
Block B is 32 mm long.
Jeremy joins some type A blocks together to make a straight row.
He then joins some type B blocks together to make a straight row of the same length.
Write down the shortest possible length of this row. **(4 marks)**

HCF and LCM page 12

You have to use a **whole number** of building blocks in each row, so the length of each row has to be a multiple of the length of one block. The answer will be the smallest number that is a multiple of 24 and a multiple of 32. This is the **lowest common multiple** of 24 and 32. You can't get all the marks just by writing down the answer. You need to show, clearly and neatly, how you got your answer.

TOP TIP

If you're not sure how to start you could draw a sketch or a scale drawing:

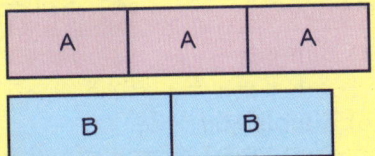

This might help you see that the lengths are multiples of 24 and 32.

21

ALGEBRA   Had a look ☐   Nearly there ☐   Nailed it! ☐

Collecting like terms

Expressions, equations and formulae

In algebra you use letters to represent unknown numbers.

$$4x + 3y - z$$

This is an **expression**. It does not have an = sign. The parts which are separated by + or − signs are called **terms**.

$$3n - 1 = 17$$

This is an **equation**. This equation only has one letter in it. You can solve an equation to find the value of the letter.

$$A = \tfrac{1}{2}bh$$

This is a **formula**. You can use it to calculate one value if you know the other values. You can't solve a formula.

Simplifying expressions

You can simplify expressions that contain + and − by **collecting like terms**.
Like terms contain the same letters.

$h + h + h = 3h$ — This means '3 lots of h' or $3 \times h$.

$5x - 2x = 3x$ — '5 lots of x' minus '2 lots of x' equals '3 lots of x'.

$2p + 3q - 5p + q = 2p - 5p + 3q + q$
$ = -3p + 4q$

$2p$ and $-5p$ are like terms. $2p - 5p = -3p$

Look at page 23 for more on simplifying expressions.

Golden rules

1. Each term includes the sign (+ or −) in front of it.
2. x means '1 lot of x'. You don't need to write $1x$
3. Like terms contain exactly the same combinations of letters with the same powers.

Worked example   Target grade 1

(a) Simplify $x + x + x + x$ (1 mark)

$4x$

(b) Simplify $n^3 + n^3$ (1 mark)

$2n^3$

Examiners' report

Practice simplifying algebraic expressions – they appear in lots of questions and if you can nail the skill you will pick up lots of marks. $n^3 + n^3$ means two lots of n^3 in total. You write this as $2n^3$.

Real students have struggled with questions like this in recent exams – **be prepared!**

Worked example   Target grade 2

Simplify $3m + 6b - 2m + b$ (2 marks)

$3m + 6b - 2m + b = 3m - 2m + 6b + b$
$ = m + 7b$

Group the m terms together and group the b terms together. Remember that each term includes the sign in front of it.

Now try this

 Target grade 1

1 (a) Simplify
     $x + x + x + x + x + x$ (1 mark)
  (b) Simplify $5q + 3q$ (1 mark)
  (c) Simplify $11t - t$ (1 mark)
  (d) Simplify $6n + n - 4n$ (1 mark)

2 | root   formula   factor   term |

Choose a word from the list above to make this sentence correct:

$8x$ is a _____ in $3y + 8x + 5$ (1 mark)

 Target grade 2

3 Simplify $8p + 2q + p - 5q$ (2 marks)

Had a look ☐   Nearly there ☐   Nailed it! ☐

ALGEBRA

Simplifying expressions

You'll need to be able to **simplify** expressions that contain × and ÷ in your exam. Use these rules to help you.

1 Multiplying expressions

1. Multiply any number parts first.
2. Then multiply the letters. Remember to use $\square^2$ for letters which are multiplied twice or $\square^3$ for letters which are multiplied three times.

$$10a \times 3a = 30a^2$$
$10 \times 3 = 30 \qquad a \times a = a^2$

$$3s \times 6t = 18st$$
$3 \times 6 = 18 \qquad s \times t = st$

2 Dividing expressions

1. Write the division as a fraction.
2. Cancel any number parts.
3. If the same letter appears on the top and bottom, you can cancel that as well.

$$8y \div 4 = \frac{{}^2\cancel{8}y}{\cancel{4}_1} = 2y \qquad 8 \div 4 = 2$$

$$\frac{{}^4\cancel{36}a\cancel{b}}{\cancel{9}_1 \cancel{b}} = 4a \qquad 36 \div 9 = 4$$

b appears on the top and the bottom, so cancel

Multiplying with algebra

You can multiply letters in algebra by writing them next to each other.
$ab = a \times b$

You can use indices to describe a letter multiplied by itself.
$y \times y = y^2$ You say 'y squared'.

You can use indices to describe the same letter multiplied together three times.
$n \times n \times n = n^3$ You say 'n cubed'.

For a reminder about squares and cubes have a look at page 8.

Worked example Target grade 3

(a) Simplify $a \times a \times a$ (1 mark)
a^3

(b) Simplify $7x \times 2y$ (2 marks)
$14xy$

(c) Simplify $10pq \div 2p$ (2 marks)
$\dfrac{{}^5\cancel{10}pq}{\cancel{2}_1 \cancel{p}} = 5q$

(b) Multiply the numbers first and then the letters.
$7 \times 2 = 14 \qquad x \times y = xy$

(c) Write the division as a fraction. You can cancel the number parts by dividing top and bottom by 2.
$10 \div 2 = 5$ so write 5 on top of your fraction.
p appears on the top and the bottom so you can cancel it. You are left with $5q$.

Now try this Target grade 3

Simplify
(a) $n \times n \times n \times n \times n$ (1 mark)
(b) $3r \times 2r \times r$ (2 marks)
(c) $4y \times 8z$ (2 marks)
(d) $20fg \div 5g$ (2 marks)
(e) $28wk \div 4wk$ (2 marks)

(d) $20fg \div 5g = \dfrac{20fg}{5g}$

Cancel any number parts then cancel any letters that are both on the top and bottom of the fraction.

23

ALGEBRA

Had a look ☐ Nearly there ☐ Nailed it! ☐

Algebraic indices

You revised indices with numbers on page 9. You can also use the index laws to simplify powers in algebraic expressions.

y^3 — This part is called the **index**. The plural of index is indices.
— This part is called the **base**.

n^5 This means $n \times n \times n \times n \times n$.

Index laws

You can use these index laws to simplify powers and algebraic expressions.

1 To multiply powers of the same base, add the indices.
$$a^m \times a^n = a^{m+n}$$
$x^4 \times x^3 = x^{4+3} = x^7$

2 To divide powers of the same base, subtract the indices.
$$a^m \div a^n = \frac{a^m}{a^n} = a^{m-n}$$
$\frac{m^8}{m^2} = m^{8-2} = m^6$

3 To raise a power of a base to a further power, multiply the indices.
$$(a^m)^n = a^{mn}$$
$(j^2)^4 = j^{2 \times 4} = j^8$

4 Powers apply to everything inside the brackets.
$$(ab)^n = a^n \times b^n$$
$(4b)^3 = 4^3 \times b^3 = 64b^3$

One at a time

When you are multiplying or dividing expressions with powers:

1. Multiply or divide any number parts first.
2. Use the index laws to work out the new power.

$7x \times 5x^6 = 35x^7$ — $x \times x^6 = x^{1+6} = x^7$

$7 \times 5 = 35$

$12 \div 3 = 4$
$\frac{12a^5}{3a^2} = 4a^3$
$a^5 \div a^2 = a^{5-2} = a^3$

(a) Add the indices. $a^m \times a^n = a^{m+n}$
(b) Subtract the indices. $a^m \div a^n = a^{m-n}$
(c) Multiply the indices. $(a^m)^n = a^{mn}$

Worked example
Target grade 4

(a) Simplify $p^2 \times p^7$ **(1 mark)**
$p^2 \times p^7 = p^{2+7} = p^9$

(b) Simplify $m^8 \div m^3$ **(1 mark)**
$m^8 \div m^3 = m^{8-3} = m^5$

(c) Simplify $(a^6)^3$ **(1 mark)**
$(a^6)^3 = a^{6 \times 3} = a^{18}$

Watch out!

You can only use the index laws when the bases are the same.

If there's no index then the number has a power of 1.

$n \times n^5 = n^{1+5} = n^6$
$x^8 \div x = x^{8-1} = x^7$

Now try this

Target grade 4

1. (a) Simplify $m^5 \times m^4$ **(1 mark)**
   (b) Simplify $k^6 \times k$ **(1 mark)**

2. (a) Simplify $h^{11} \div h^3$ **(1 mark)**
   (b) Simplify $\frac{t^{25}}{t^5}$ **(1 mark)**
   (c) Simplify $\frac{a^3 \times a^6}{a^4}$ **(1 mark)**

3. (a) Simplify $(y^4)^2$ **(1 mark)**

Target grade 5
   (b) Simplify $(5p^2)^3$ **(1 mark)**

You could do this in two different ways:
$(5p^2)^3 = 5^3 \times (p^2)^3$
$(5p^2)^3 = 5p^2 \times 5p^2 \times 5p^2$
Both ways will give you the same answer.

Worked solution video

Had a look ☐   Nearly there ☐   Nailed it! ☐

ALGEBRA

Substitution

If you know the values of the letters in an algebraic expression, you can **substitute** them into the expression. This lets you work out the value of the expression.

$x = 7$ and $y = 2$ have been substituted into this expression.

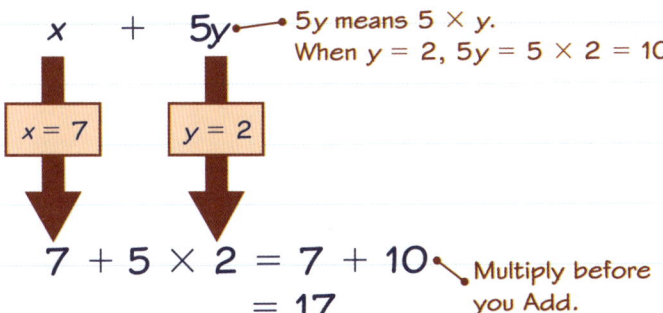

When $x = 7$ and $y = 2$ the value of $x + 5y$ is 17.

Using BIDMAS

Remember to use the correct priority of operations when you are doing a calculation. BIDMAS gives the priority in which the operations should be carried out.

Brackets
Indices
Division
Multiplication
Addition
Subtraction

$2 \times 5^2 - (14 + 8)$
$= 2 \times 5^2 - 22$
$= 2 \times 25 - 22$
$= 50 - 22$
$= 28$

You should substitute all the values before doing any calculations.

Worked example — Target grade 3

(a) Work out the value of $5x + 1$ when $x = -3$ **(2 marks)**

$5 \times (-3) + 1 = -15 + 1$
$= -14$

(b) Work out the value of $5p^3$ when $p = 2$ **(2 marks)**

$5 \times 2^3 = 5 \times 8$
$= 40$

(c) Work out the value of $3m + 4n$ when $m = 5$ and $n = -2$ **(2 marks)**

$3 \times 5 + 4 \times (-2) = 15 + (-8)$
$= 15 - 8$
$= 7$

(d) Work out the value of $2x(x - 1)$ when $x = 11$ **(2 marks)**

$2 \times 11 \times (11 - 1) = 2 \times 11 \times 10$
$= 22 \times 10$
$= 220$

Examiners' report

Be extra careful when substituting a negative number. You can use brackets around the number to make sure you don't make a mistake.

Real students have struggled with questions like this in recent exams – **be prepared!**

Remember **BIDMAS**. Indices comes before Multiplication.

Substitute all the values before starting your calculation.

You can multiply in any order. $22 \times 10 = 2 \times 110$

Now try this — Target grade 3

1 Work out the value of $h + k - 6n$ when $h = 5$, $k = 10$ and $n = \frac{1}{2}$ **(2 marks)**

2 Work out the value of $\dfrac{xw}{x + y}$ when $x = -8$, $w = 3$ and $y = 4$ **(3 marks)**

First work out what the top and bottom of the fraction come to, then work out the division.

3 $x = 3$, $y = 6$ and $z = 10$
Work out the value of these expressions.
(a) $x + z$ **(1 mark)**
(b) $yz - x^2$ **(1 mark)**
(c) $x^3 - (z - y)$ **(1 mark)**

Worked solution video

| ALGEBRA | Had a look ☐ | Nearly there ☐ | Nailed it! ☐ |

Formulae

A **formula** is a mathematical rule.
Formulae is the plural of formula.
The formula for the area of this triangle is:
 Area = $\frac{1}{2}$ × Base × Vertical height
You can write this formula using algebra as:
 $A = \frac{1}{2}bh$

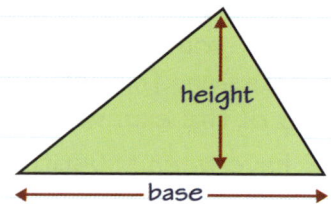

A formula lets you calculate one quantity when you know the others.
You need to **substitute** the values you know into the formula.

For more on substitution have a look at page 25.

Worked example — Target grade 1

You can use this formula to work out the cooking time in minutes for a turkey.

Cooking time = Weight in kg × 30 + 45

Work out the cooking time for a turkey weighing 7 kg. **(3 marks)**

Cooking time = 7 × 30 + 45
 = 210 + 45
 = 255

The cooking time is 255 minutes, or 4 hours and 15 minutes.

Substitute the weight into your formula before you do any calculations.

Remember to use **BIDMAS** for the correct order of operations. You **M**ultiply before you **A**dd.

For a reminder about substituting and BIDMAS have a look at page 25.

When you are giving an answer in minutes and it is larger than 60 minutes, you can give it in minutes or in hours and minutes.
4 × 60 = 240 and 255 − 240 = 15
So 255 minutes = 4 hours and 15 minutes.

Substitute the values for u and t into the formula. Using **BIDMAS**:

1. Do the **I**ndex (power) first.
 $3^2 = 9$

2. Do the **M**ultiplications next.
 20 × 3 = 60
 5 × 9 = 45

3. Do the **S**ubtraction last.
 60 − 45 = 15

Don't try to do more than one operation on each line of working.

Worked example — Target grade 3

This formula is used in physics to calculate distance.
$D = ut - 5t^2$
$u = 20$
$t = 3$
Work out the value of D. **(3 marks)**

$D = 20 × 3 - 5 × 3^2$
 $= 20 × 3 - 5 × 9$
 $= 60 - 45$
 $= 15$

Now try this

1 Here is a trapezium.
 The area of the trapezium is given by the formula Area = $\frac{1}{2}(a + b)h$
 Work out the area when
 $h = 4.5$, $a = 6.5$ and $b = 9.5$

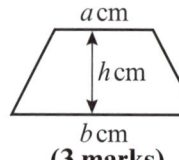

(3 marks)

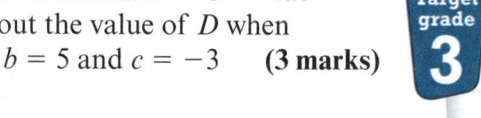

2 Here is a formula: $D = b^2 - 4ac$
 Work out the value of D when
 $a = 2$, $b = 5$ and $c = -3$ **(3 marks)**

Substitute all the values before you do any calculations.

Had a look ☐ Nearly there ☐ Nailed it! ☐

ALGEBRA

Writing formulae

You can write a rule given in words as a word formula or as a formula using algebra.

This label gives instructions for working out the cooking time of a chicken.

FREE-RANGE CHICKEN		
WEIGHT (KG)	PRICE PER KG	COOKING INSTRUCTIONS
1.8	£3.95	Cook at 170°C for 25 minutes per kg plus half an hour

You can write the cooking instructions as a word formula.

Cooking time in minutes = 25 × Weight in kg + 30

You need to give units when you are describing the quantities in a formula. If the cooking time is in hours then this formula would give you a very crispy chicken!

You can also write this formula using algebra.

$T = 25w + 30$, where T is the cooking time in minutes and w is the weight in kg.

When you write a formula using algebra you need to explain what each letter means.

Worked example — Target grade 3

Chloe buys a pens costing 25 pence each and b pencils costing 15 pence each. Write a formula for the total cost T pence. **(2 marks)**

$T = 25a + 15b$

$25a$ means $25 \times a$. You should make sure your formula is simplified as much as possible. Don't write any units (like pence) in your formula.

Worked example — Target grade 3

The cost of hiring a car can be worked out using this formula.

Cost = £80 + 50p per mile

Write a formula for the cost £C of hiring a car which is driven for m miles. **(2 marks)**

$C = 80 + 0.5m$

You need to make sure all your values are in the same units. C is the cost in pounds, so convert 50p into pounds. 50p = £0.50, so you need to multiply the number of miles by 0.5

Worked example — Target grade 3

The diagram shows a regular hexagon.

Write a formula for the perimeter of the hexagon P in terms of s. **(1 mark)**

$P = 6s$

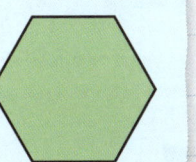

The perimeter is the distance all the way around a shape.

All the sides of a regular shape are the same length. There are 6 sides so:

Perimeter = 6 × length of one side

$P = 6s$

You can find out more about perimeter on page 79.

Now try this

Target grade 2

1. Rulers cost 50 pence each and pens cost 35 pence each.
   (a) Write an expression for the total cost in pence of x rulers and y pens. **(1 mark)**
   (b) Complete this formula for the total cost, T pence.

   $T = $ _____ **(1 mark)**

Target grade 3

2. This shape is made from 3 identical squares.

   (a) Write a formula for the area, A, of the shape. **(1 mark)**
   (b) Work out the area of the shape when $s = 5$ cm. **(1 mark)**

Worked solution video

ALGEBRA Had a look ☐ Nearly there ☐ Nailed it! ☐

Expanding brackets

Expanding brackets is sometimes called **multiplying out** brackets.

> **Golden rule**
> You have to multiply the expression outside the brackets by everything inside the brackets.
>
> $4n \times n = 4n^2$
>
> $4n(n + 2) = 4n^2 + 8n$
>
> $4n \times 2 = 8n$
>
> For a reminder about multiplying expressions look back at page 23.

You need to be extra careful if there are negative signs outside the brackets.

$-2 \times x = -2x$

$-2(x - y) = -2x + 2y$

$-2 \times -y = 2y$

Multiplying negative terms is like multiplying negative numbers.

When both terms are + or both terms are − the answer is **positive**.
$-2a \times -a = +2a^2$

When one term is + and one term is − the answer is **negative**.
$-10p \times 5q = -50pq$

Sometimes you have to **expand and simplify**. This means 'multiply out the brackets and then collect like terms'.

$6 \times m = 6m \qquad 4 \times 3 = 12$

$6(m + 2) + 4(3 - m) = 6m + 12 + 12 - 4m$

$6 \times 2 = 12 \quad 4 \times -m = -4m \quad = 2m + 24$

$6m - 4m = 2m \qquad 12 + 12 = 24$

Remember that any negative signs belong to the term on their right.
Look back at page 22 for a reminder about collecting like terms.

Worked example *Target grade 4*

(a) Expand $5(2y - 3)$ **(2 marks)**

$5(2y - 3) = 10y - 15$

(b) Expand and simplify $2(3x + 4) - 3(4x - 5)$ **(3 marks)**

$2(3x + 4) - 3(4x - 5)$
$= 6x + 8 - 12x + 15$
$= -6x + 23$

> For part (b), be careful with the second bracket.
> $-3 \times 4x = -12x$
> $-3 \times -5 = +15$
> The question says 'expand and simplify' so remember to collect any like terms after you have multiplied out the brackets.

Now try this *Target grade 4*

1 Expand (a) $3(y - 6)$ **(2 marks)** (b) $m(m + 7)$ **(2 marks)**

2 Expand and simplify
   (a) $5(a - 2b) + 4(2a + b)$ **(3 marks)**
   (b) $2(4w + 3) - 3(3w - 5)$ **(3 marks)**
   (c) $2m(m + 9) - m(m - 4)$ **(3 marks)**

> Be careful when there is a minus sign in front of the second bracket.
> $-3 \times 3w = -9w$ and $-3 \times -5 = 15$

Had a look ☐   Nearly there ☐   Nailed it! ☐   **ALGEBRA**

Factorising

Factorising is the opposite of expanding brackets.

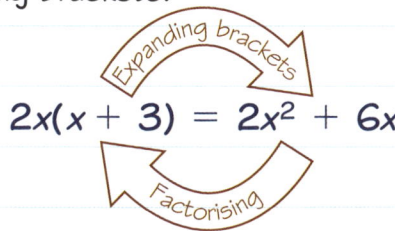

$2x(x + 3) = 2x^2 + 6x$

You need to look for the largest factor you can take out of every term in the expression. This is what you need to write outside the brackets.

You have to factorise expressions as much as possible.

$10a^2 + 5a = 5(2a^2 + a)$
This expression has only been **partly factorised**.

$10a^2 + 5a = 5a(2a + 1)$
This expression has been **completely factorised**. The two **factors** are $5a$ and $(2a + 1)$

To factorise ...

1. Look for the **largest factor** you can take out of every term.
   $12x - 8$
   The largest factor is 4.

2. Write this factor outside the brackets. Write the sign (+ or −) from the expression inside the brackets.
   $12x - 8 = 4(\quad - \quad)$

3. Work out what you need to multiply the factor by to get each term in the original expression.
   $12x - 8 = 4(3x - 2)$

4. **Check** your answer by expanding the brackets. You should get back to the original expression.

   $4 \times 3x = 12x$
   $4(3x - 2) = 12x - 8$ ✓
   $4 \times -2 = -8$

Examiners' report

When you write an expression with brackets both parts of the factorised expression are factors:
$6x^2 + 12xy = 6x(\underline{x + 2y}) = \underline{3x}(2x + 4y)$
$\qquad = \underline{x}(6x + 12y)$
So the three underlined expressions above are all factors of the original expression.

Real students have struggled with questions like this in recent exams – **be prepared!**

Worked example   Target grade 5

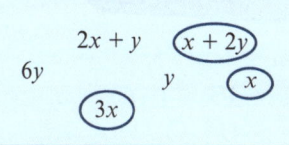
$6y \qquad 3x$

Circle all the expressions in the box above that are factors of $6x^2 + 12xy$ **(2 marks)**

Worked example   Target grade 4

(a) Factorise $y^2 + 3y$ **(1 mark)**
$y^2 + 3y = y(y + 3)$

(b) Factorise fully $2p^2 - 4p$ **(2 marks)**
$2p^2 - 4p = 2p(p - 2)$

(a) **Check it!**
$y \times y = y^2$ ✓ $y \times 3 = 3y$ ✓

(b) The question says 'factorise fully'. This means you have to take the largest factor outside the brackets.
$2p^2 - 4p = 2(p^2 - 2p)$ is not fully factorised.

Now try this

Target grade 3
1 Factorise
 (a) $6n + 18$ **(1 mark)**
 (b) $3p - 3$ **(1 mark)**

Target grade 4
2 Factorise fully
 (a) $4t^2 + 8t$ **(2 marks)**
 (b) $3x^2 - 12x$ **(2 marks)**

Target grade 5
3 | 2   4p   2p   p + 2q   p − 2q |

Write down the factors of $2p^2 - 4pq$ from the list.   **(2 marks)**

Worked solution video

ALGEBRA

Had a look ☐   Nearly there ☐   Nailed it! ☐

Linear equations 1

An **equation** is like a pair of scales. The equals sign tells you that the scales are **balanced**. The letter represents an unknown weight. You can solve the equation to find the value of the letter.

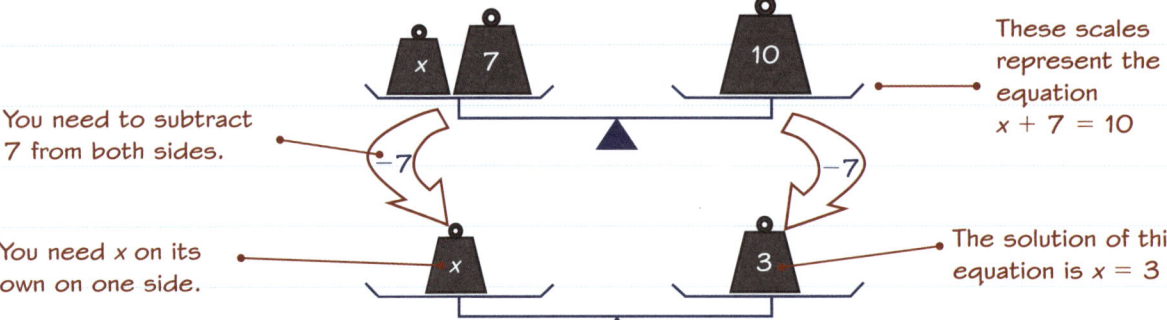

You need to subtract 7 from both sides.

These scales represent the equation $x + 7 = 10$

You need x on its own on one side.

The solution of this equation is $x = 3$

As long as you do the same thing to both sides, the scales stay balanced.

To **solve** an equation you need to get the letter on its own on one side.

It is important to write your working neatly when you are solving equations.

Every line of working should have an $=$ sign in it.

Start a new line for each step. Do one operation at a time.

$$5x + 3 = 18 \quad (-3)$$
$$5x = 15 \quad (\div 5)$$
$$x = 3$$

This is the solution to the equation. Your final line of working should look like this.

Write down the operation you are carrying out. Remember to do the same thing to both sides of the equation.

$5x$ means $5 \times x$. You have to divide by 5 to get x on its own.

Worked example — Target grade 1

(a) Solve $m + 6 = 15$ **(1 mark)**

$m + 6 = 15 \quad (-6)$
$m = 9$

(b) Solve $8p = 36$ **(1 mark)**

$8p = 36 \quad (\div 8)$
$p = 4.5$

In part (b) you need to do **two operations** to get a on its own. Do them one at a time, and make sure you show your working.

Worked example — Target grade 2

(a) Solve $\frac{x}{3} = -5$ **(1 mark)**

$\frac{x}{3} = -5 \quad (\times 3)$
$x = -15$

(b) Solve $2a - 7 = 11$ **(2 marks)**

$2a - 7 = 11 \quad (+7)$
$2a = 18 \quad (\div 2)$
$a = 9$

Now try this

1 Solve
  (a) $3y = 24$ **(1 mark)**
  (b) $m + 8 = 25$ **(1 mark)**
  (c) $31 - k = 12$ **(1 mark)**

2 Solve
  (a) $4n - 7 = 13$ **(2 marks)**
  (b) $2w + 11 = 32$ **(2 marks)**

3 (a) Solve $\frac{n}{4} = 9$ **(1 mark)**

  (b) Solve $\frac{q}{11} = 7$ **(1 mark)**

Your answer doesn't have to be a whole number.

Worked solution video

Had a look ☐ Nearly there ☐ Nailed it! ☐ **ALGEBRA**

Tricky Topic

Linear equations 2

Equations with brackets

Always start by multiplying out the brackets then collecting like terms.
For a reminder about multiplying out brackets have a look at page 28.

$2 \times 3y = 6y$

$2(3y + 5)$

$2 \times 5 = 10$

$2(3y + 5) = 22$
$6y + 10 = 22$ (-10)
$6y = 12$ $(\div 6)$
$y = 2$

Do one operation at a time. Write down the operation you are using at each step.

Equations with the letter on both sides

To solve an equation you have to get the letter on its own on one side of the equation.
Start by collecting like terms so that all the letters are together.

You can add or subtract multiples of x on both sides of the equation.

$4x + 26 = 2 - 2x$ $(+ 2x)$
$6x + 26 = 2$ $(- 26)$
$6x = -24$ $(\div 6)$
$x = -4$

Remember to do the same thing to both sides of the equation.

Worked example

 Target grade 4

Solve $7p + 2 = 5(p - 4)$ **(3 marks)**

$7p + 2 = 5(p - 4)$ (expand brackets)
$7p + 2 = 5p - 20$ $(- 5p)$
$2p + 2 = -20$ $(- 2)$
$2p = -22$ $(\div 2)$
$p = -11$

Start by expanding the brackets on the right-hand side. Then subtract $5p$ from both sides to get all the p terms together.

Examiners' report

Don't use a trial method to solve an equation. You probably won't find the correct answer, and you can't get any method marks.

Real students have struggled with questions like this in recent exams – **be prepared!**

Worked example

 Target grade 4

Solve $\dfrac{10 - y}{5} = 3$ **(3 marks)**

$\dfrac{10 - y}{5} = 3$ $(\times 5)$

$10 - y = 15$ $(+ y)$
$10 = 15 + y$ (-15)
$-5 = y$

Start by multiplying both sides by 5 to remove the fraction. You want the y term to be positive so add y to both sides of the equation.

You can write your answer in the form $-5 = y$. As long as the letter is on its own on one side then the equation is solved.

Now try this

 Target grade 3

1 (a) Solve $33 - 5x = 8$ **(2 marks)**
   (b) Solve $4 - 6x = 5 - 8x$ **(3 marks)**

A **positive** x term is easier to work with. Start by adding $5x$ to both sides.

2 (a) Solve $3t - 7 = 5t + 15$ **(3 marks)**
   (b) Solve $2(4h + 3) = 2$ **(3 marks)**

 Target grade 4

3 (a) Solve $3y + 19 = 5(y - 2)$ **(3 marks)**
   (b) Solve $\dfrac{16 - 5m}{3} = 7$ **(3 marks)**

Expand the brackets then group the y terms on one side.

Worked solution video

31

ALGEBRA

Had a look ☐   Nearly there ☐   Nailed it! ☐

Inequalities

You can use these symbols to describe **inequalities**. The fat end of the symbol always points towards the bigger number.

- $>$ means 'is greater than'
- $\geq$ means 'is greater than or equal to'
- $<$ means 'is less than'
- $\leq$ means 'is less than or equal to'

Error intervals

You can use inequalities to show the error interval when a number has been **rounded**:

1 The height of the London Eye is 140 m (to the nearest 10 m). You can write
$135\,m \leq \text{Height} < 145\,m$

The least possible value for the height is 135 m. — You use 'less than' for the greatest possible value.

2 $x = 4.7$ (1 d.p.) means $4.65 \leq x < 4.75$

You use 'less than or equal to' for the least possible value.

Worked example   Target grade 4

Aron rounds a number, n, to 2 decimal places. The result is 8.64.
Using inequalities, write down the error interval for n. **(2 marks)**

$8.635 \leq n < 8.645$

Inequalities on number lines

You can represent inequalities on a number line.

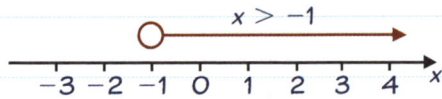   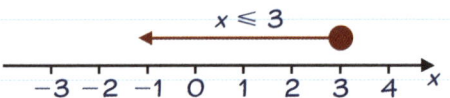

The open circle shows that -1 is **not** included.   The closed circle shows that 3 **is** included.

Worked example   Target grade 4

(a) Show the inequality $-1 < x \leq 4$ on the number line below.

(1 mark)

(b) x is an integer and $-1 < x \leq 4$. Write down all the possible values of x.

0, 1, 2, 3, 4   **(2 marks)**

Examiners' report

Integers are positive and negative whole numbers, including zero. So you **should** include zero in your answer to part (b). Look at the inequalities carefully.

- The bottom inequality is $<$ so **don't** include -1.
- The top inequality is $\leq$ so **do** include 4.

Real students have struggled with questions like this in recent exams – **be prepared!**

Now try this   Target grade 4

1 Write down the inequalities shown on these number lines.

(a) 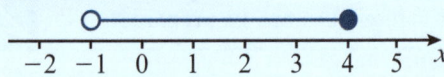   **(1 mark)**

(b) 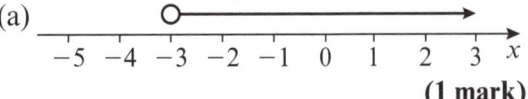   **(1 mark)**

2 Show these inequalities on the number lines below.

(a) $-4 < x < 1$ **(1 mark)**   (b) $-2 \leq x < 5$ **(1 mark)**

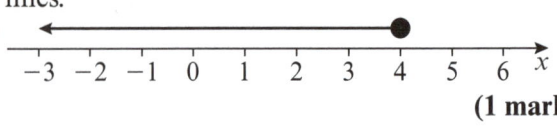

Had a look ☐   Nearly there ☐   Nailed it! ☐   **ALGEBRA**

Solving inequalities

You can solve an inequality in exactly the same way as you solve an equation.

$2x - 3 \leq 15$ $(+ 3)$
   $2x \leq 18$ $(\div 2)$
     $x \leq 9$

Solutions
The solution has the letter on its own on one side of the inequality and a number on the other side.

Solutions to inequalities
$x > 4$ $-2 < x$ $x \leq -\frac{3}{4}$

Not solutions to inequalities
$x \geq 20 + 3$ $2x < 10$ $x = 4$

Watch out!
If you multiply or divide by a **negative number** when solving an inequality, you have to **reverse the direction** of the inequality sign. It's usually easier to eliminate negative terms by **adding**.

Worked example — Target grade 4

Solve the inequality
(a) $2x + 1 < 10$ $(- 1)$ **(2 marks)**
     $2x < 9$ $(\div 2)$
     $x < \frac{9}{2}$ or $x < 4\frac{1}{2}$

(b) $5x < 2x - 6$ $(- 2x)$ **(2 marks)**
     $3x < -6$ $(\div 3)$
     $x < -2$

(c) $12 - 4x \geq 10$ $(+ 4x)$ **(3 marks)**
     $12 \geq 10 + 4x$ $(- 10)$
     $2 \geq 4x$ $(\div 4)$
     $\frac{2}{4} \geq x$ or $\frac{1}{2} \geq x$

These are **inequalities** and not equations. You don't need to use an = sign anywhere in your answer. Write down the operation you are using at each step and remember you have to do the same thing to both sides of the inequality.
Your answer can involve negative numbers or fractions.

When you're solving an inequality you should avoid multiplying or dividing by a **negative** number. Add 4x to both sides of the equation to make the x term positive.

Worked example — Target grade 4

Write down all the integers that satisfy the inequality:
$-15 \leq 5n < 10$ **(3 marks)**

$-15 \leq 5n$ $5n < 10$ $(\div 5)$
$-3 \leq n$ $n < 2$
$n = -3, -2, -1, 0, 1$

This is really two inequalities:
$-15 \leq 5n$ and $5n < 10$
Solve them separately, then write down all the integers which satisfy **both** inequalities at the same time.

Now try this — Target grade 4

1. Solve the following inequalities.
   (a) $4x + 3 \geq 23$ **(2 marks)**
   (b) $18 - 5x > -2$ **(3 marks)**

2. Find all the integers which satisfy each of these inequalities.
   (a) $-8 \leq 4x < 12$ **(3 marks)**
   (b) $-1 < 3x \leq 10$ **(3 marks)**

Worked solution video

ALGEBRA   Had a look ☐   Nearly there ☐   Nailed it! ☐

Sequences 1

A **sequence** is a pattern of numbers or shapes that follow a rule.

2, 4, 6, 8, 10... is a sequence of even numbers.

1, 4, 9, 16, 25... is a sequence of square numbers.

Each number in a number sequence is called a **term**.

You can continue a sequence of numbers by finding the rule to get from one term to the next.

Worked example (Target grade 2)

The numbers in this sequence increase by the same amount each time:

11   <u>15</u>   19   23   27   <u>31</u>
            +4 +4

Work out the two missing numbers. **(2 marks)**

Everything in blue is part of the answer.

Problem solved! You are given the third term. To work out the term **before** this you need to work **backwards** through the rule. You can use function machines to help you.

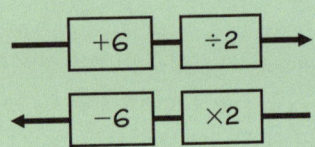

Check it!

$(14 + 6) \div 2 = 20 \div 2 = 10$

$(10 + 6) \div 2 = 16 \div 2 = 8$ ✓

Worked example (Target grade 3)

Here is a sequence. The third term of this sequence is 8.

...   ...   8   ...

The rule for this sequence is

| Add 6 to the previous term and then divide by 2 |

Work out the first term of this sequence.

(3 marks)

3rd term = 8
2nd term = 8 × 2 − 6 = 10
1st term = 10 × 2 − 6 = 14

Generating sequences

You can work out the terms of a sequence by substituting the term number into the nth term. Here are some examples:

nth term	$9 - 2n$	$n^2 + 10$
1st term	$9 - 2 \times 1 = 7$	$1^2 + 10 = 11$
2nd term	$9 - 2 \times 2 = 5$	$2^2 + 10 = 14$
↓	↓	↓
8th term	$9 - 2 \times 8 = -7$	$8^2 + 10 = 74$

Worked example (Target grade 3)

The nth term of a sequence is $50 - n^2$

Work out the first term of the sequence that is negative. **(2 marks)**

7th term = $50 - 7^2 = 1$
8th term = $50 - 8^2 = -14$
The 8th term is the first negative term.

Now try this

 1 The first five terms of a sequence are
   3 5 9 15 23
   Write down the next two terms of this sequence.
   (1 mark)

 2 The nth term of a sequence is $7n - 5$
   (a) Work out the 8th term of this sequence.
   (1 mark)
   (b) Work out the first term of this sequence that is greater than 70.   **(1 mark)**

 3 The rule for generating a sequence is 'add two consecutive terms to get the next term'.
   Here are the first five terms:
   2 3 5 8 13
   Write down the next three terms of this sequence.   **(1 mark)**

*This type of sequence is called a **Fibonacci-type** sequence.*

Had a look ☐   Nearly there ☐   Nailed it! ☐   **ALGEBRA**

Sequences 2

An **arithmetic** or **linear sequence** is a sequence of numbers where the difference between consecutive terms is **constant**. In your exam, you might need to work out the *n*th term of a sequence. Look at this example which shows you how to do it in four steps.

Worked example

Target grade 4

1 Here is a sequence:

1 +4→ 5 +4→ 9 +4→ 13 +4→ 17

Work out a formula for the *n*th term of the sequence. **(2 marks)**

Write in the difference between each term.

2 Here is a sequence:
Zero term
−3   1 +4→ 5 +4→ 9 +4→ 13 +4→ 17

Work out a formula for the *n*th term of the sequence. **(2 marks)**

*Work backwards to find the **zero term** of the sequence. You need to subtract 4 from the first term.*

3 Here is a sequence:
Zero term
−3   1 +4→ 5 +4→ 9 +4→ 13 +4→ 17

Work out a formula for the *n*th term of the sequence. **(2 marks)**

*n*th term = Difference × *n* + Zero term

*Write down the formula for the nth term. **Remember** this formula for the exam.*

4 Here is a sequence:
Zero term
−3   1 +4→ 5 +4→ 9 +4→ 13 +4→ 17

Work out a formula for the *n*th term of the sequence. **(2 marks)**

*n*th term = Difference × *n* + Zero term

*n*th term = 4*n* − 3

Is 99 in this sequence?

You can use the *n*th term to check whether a number is a term in the sequence.

The value of *n* in your *n*th term has to be a **positive** whole number.

Try some different values of *n*:

$n = 25 \rightarrow 4n - 3 = 97$
$n = 26 \rightarrow 4n - 3 = 101$

You can't use a value of *n* between 25 and 26 so 99 is **not** a term in the sequence.

Check it!

Check your answer by substituting values of *n* into your *n*th term.
1st term: when $n = 1$,
$4n - 3 = 4 \times 1 - 3 = 1$ ✓
2nd term: when $n = 2$,
$4n - 3 = 4 \times 2 - 3 = 5$ ✓

You can also generate any term of the sequence.

For the 20th term: when $n = 20$,
$4n - 3 = 4 \times 20 - 3 = 77$

So the 20th term is 77.

Now try this

Target grade 4

Here are the first five terms of an arithmetic sequence:
3   7   11   15   19

(a) Write down an expression for the *n*th term. **(2 marks)**

Karen says that 89 is a term in the sequence.

(b) Is she right? Give reasons for your answer. **(2 marks)**

Write down some terms in the sequence that are close to 89. Remember to write a conclusion to answer the question.

| ALGEBRA | Had a look ☐ | Nearly there ☐ | Nailed it! ☐ |

Coordinates

You can use coordinates to describe the positions of points on a grid.

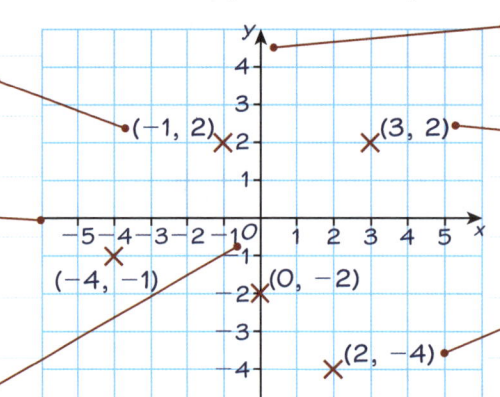

- The vertical axis is labelled y.
- The first number in a coordinate pair describes the horizontal position. The second number describes the vertical position.
- You can use negative numbers to describe points below 0 on the y-axis.

This point is to the left of 0 on the x-axis. So the x-coordinate is negative.

The horizontal axis is labelled x.

The point O is called the origin and has coordinates (0, 0).

Worked example

Target grade 1

(a) Write down the coordinates
 (i) of the point A (1 mark)
 (2, 3)
 (ii) of the point B. (1 mark)
 (2, −2)

(b) On the grid, plot the point D so that ABCD is a rectangle. (1 mark)

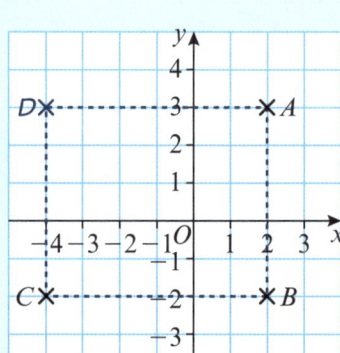

Everything in blue is part of the answer.

A rectangle has four right angles. Plot point D level with point A and vertically above point C.

Mid-points

A **line segment** is a short section of a straight line.
The mid-point of a line segment is exactly halfway along the line. You can find the mid-point if you know the coordinates of the ends.

To find the mid-point, add the x-coordinates and divide by 2 and add the y-coordinates and divide by 2.

$$\text{Mid-point} = \left(\frac{x_1 + x_2}{2}, \frac{y_1 + y_2}{2}\right)$$

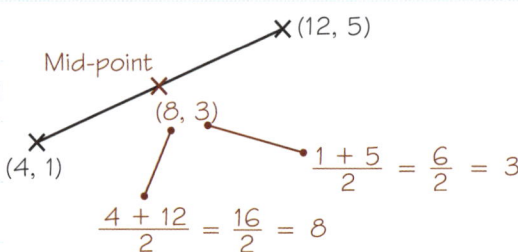

$\frac{4 + 12}{2} = \frac{16}{2} = 8$

$\frac{1 + 5}{2} = \frac{6}{2} = 3$

You need to be confident in finding mid-points if you're aiming for a top grade.

Now try this

Target grade 3

ABCD is a rectangle.
(a) Work out the coordinates of D. (1 mark)
(b) Work out the length of AB. (1 mark)
(c) E is the centre of the rectangle.
 Work out the coordinates of E. (2 marks)

E is the mid-point of AC and of BD.

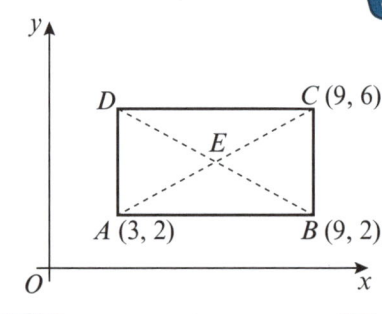

Had a look ☐ Nearly there ☐ Nailed it! ☐ **ALGEBRA**

Gradients of lines

The **gradient** of a straight-line graph measures how steep the line is. You can work out the gradient by drawing a triangle and using this rule: Gradient = $\dfrac{\text{Distance up}}{\text{Distance across}}$

Worked example

Target grade 3

This scatter graph shows the relationship between the budgets of some films and the amounts of money they made at the box office in their opening weekends. A line of best fit has been drawn on the scatter graph.

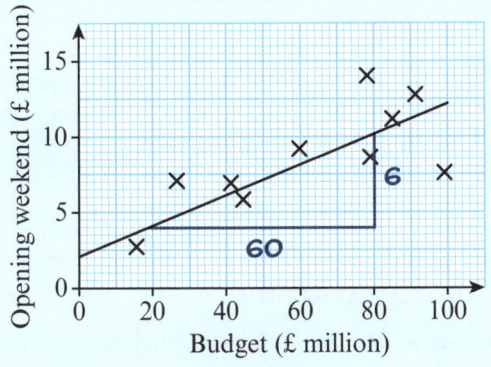

Everything in blue is part of the answer.

(a) Work out the gradient of the line of best fit. **(1 mark)**

Gradient = $\dfrac{\text{Distance up}}{\text{Distance across}} = \dfrac{6}{60} = 0.1$

(b) Amir says, 'Spending an extra £10 million on a film is likely to increase the amount of money taken during its opening weekend by £1 million.' Does the graph support this statement? **(1 mark)**

Yes. The gradient of the line of best fit is 0.1, so the amount of money taken during the opening weekend increases by approximately £0.1 million for each additional £1 million of budget.

To work out the gradient of a line you need to draw a triangle.

Write the distance across and the distance up.

Watch out for the scales on the axes:

Distance across = 80 − 20 = 60

Distance up = 10 − 4 = 6

There is more on scatter graphs on page 119.

Top triangle tips!

1. Draw one side of your triangle on a large grid line as you are less likely to make a mistake in your calculations.
2. Use a large triangle as this means your calculations are more accurate.
3. Don't just count grid squares. Use the scale to work out the distance across and the distance up.

Positive or negative?

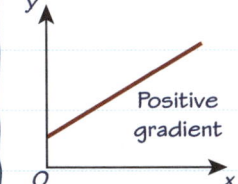 Positive gradient

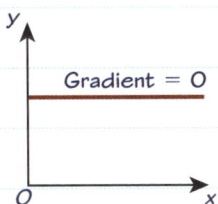 Gradient = 0

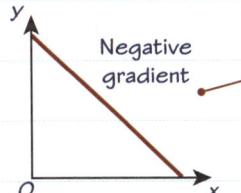 Negative gradient

If the gradient is negative then one value decreases as the other value increases.

Now try this

Target grade 3

Use the **scale** when calculating the gradient.

The scatter graph shows a relationship between the amount of money a company spent on marketing each month and its revenue.

(a) Calculate the gradient of the line of best fit.
(b) Amy states that 'For every £10 000 spent on advertising, revenue increases by £20 000.'
Does the scatter graph support her statement? **(1 mark)**

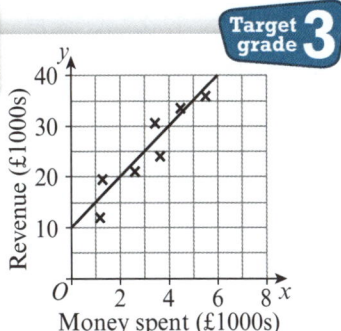

ALGEBRA Had a look ☐ Nearly there ☐ Nailed it! ☐

Straight-line graphs 1

Here are two things you need to know about straight-line graphs:

1 If an equation is in the form $y = mx + c$, its graph will be a straight line.

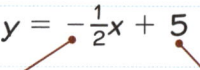

This number tells you the gradient of the graph.

The y-intercept of the graph is at $(0, 5)$.

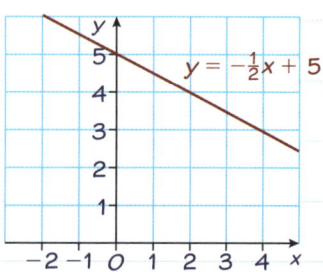

The gradient of the graph is $-\frac{1}{2}$. This means that for every unit you go across, you go half a unit down.

2 Use a table of values to draw a graph.
$y = 2x + 1$

x	−1	0	1	2
y	−1	1	3	5

  $y = 2 \times 2 + 1 = 5$

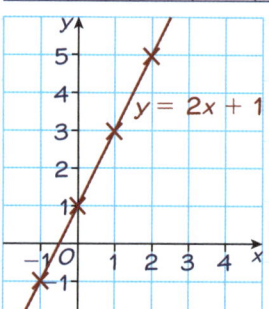

Choose simple values of x and substitute them into the equation to find the values of y.

Plot the points on your graph and join them with a straight line.

Worked example *Target grade 4*

On the grid draw the graph of $x + y = 4$ for values of x from -2 to 5.

(3 marks)

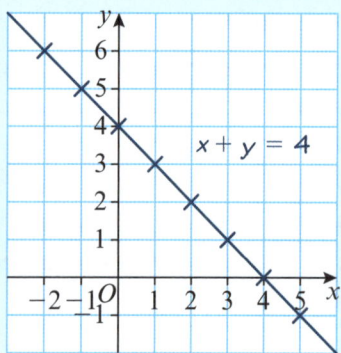

Everything in blue is part of the answer.

x	−2	−1	0	1	2	3	4	5
y	6	5	4	3	2	1	0	−1

You can rearrange the equation of this graph into the form $y = mx + c$ so it is a **straight line**.

$x + y = 4$ $(-x)$
$y = -x + 4$ $m = -1$ and $c = 4$

The gradient is -1 and the y-intercept is at $(0, 4)$. You could use this information to draw the graph, but it's safer to make a table of values. Make sure you plot **at least three** points, then join them with a straight line **using a ruler**.

Finding equations

If you have a graph you can find its equation by working out the gradient and looking at the y-intercept.

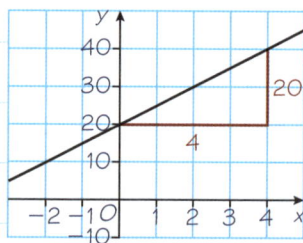

Draw a triangle to find the gradient.

Gradient $= \frac{20}{4} = 5$

The y-intercept is $(0, 20)$.

Put your values for gradient, m, and y-intercept, c, into the equation of a straight line, $y = mx + c$.

The equation is $y = 5x + 20$

Now try this *Target grade 4*

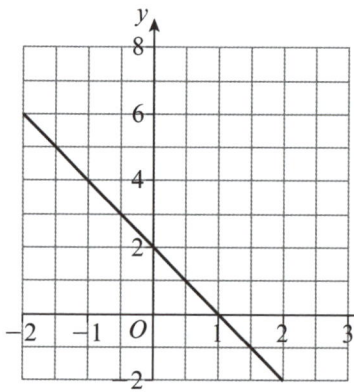

Find the equation of the straight line.

(3 marks)

Use $y = mx + c$. Draw a triangle to find the gradient, m. The graph slopes down so m will be **negative**.

Had a look ☐   Nearly there ☐   Nailed it! ☐

ALGERBA

Straight-line graphs 2

You can use **algebra** to find the equation of the line when you are given one point and the gradient, or two points.

1 Given one point and the gradient

| Substitute the gradient for m in $y = mx + c$ | Gradient 2, passing through point (3, 7) |

↓

| Substitute the x- and y-values given into the equation | $y = 2x + c$
$7 = 2 \times 3 + c$
$7 = 6 + c$ (-6)
$c = 1$ |

↓

| Solve the equation to find c | $y = 2x + 1$ |

↓

| Write out the equation |

2 Given two points

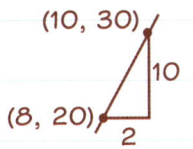

| Draw a sketch showing the two points | Passing through points (8, 20) and (10, 30) |

↓

| Work out the gradient of the line with a triangle | |

↓

| Use method 1 (on the left) and one of the points given to find the equation | Gradient $= \dfrac{10}{2} = 5$
$y = 5x + c$
$30 = 5 \times 10 + c$
so $c = -20$
$y = 5x - 20$ |

Parallel lines

Parallel lines have the same gradient. These three lines all have a gradient of 1.

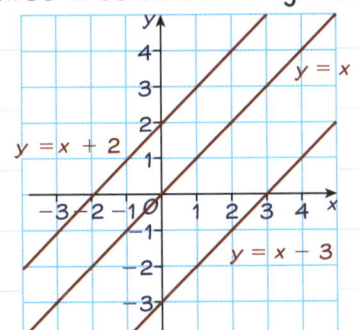

Parallel lines **never meet**.

Worked example   Target grade 5

Line A has equation $y = \frac{1}{2}x + 1$
Line B is parallel to line A.

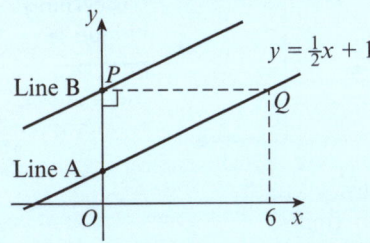

Work out the equation of line B.   **(3 marks)**

$y = \frac{1}{2}x + 1$
$= \frac{1}{2} \times 6 + 1 = 3 + 1 = 4$

So Q has coordinates (6, 4)
So P has coordinates (0, 4)
Line B is parallel to line A so also has gradient $\frac{1}{2}$
So equation of line B is $y = \frac{1}{2}x + 4$

 Plan your strategy before you start:

1. Use the equation of line A to find the coordinates of points P and Q.
2. Use the fact that line A is parallel to line B to write down the gradient of line B.
3. Use $y = mx + c$ to write down the equation of line B.

Now try this

 Target grade 5

1 A straight line has gradient 6 and passes through the point (5, 10). Find the equation of the line.   **(2 marks)**

2 Find the equation of the straight line which passes through the points (3, 5) and (7, 13).   **(3 marks)**

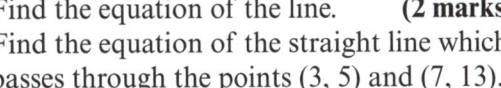

Find the gradient, m, then substitute $x = 3$ and $y = 5$ into $y = mx + c$. Solve the equation to find the value of c.

39

ALGEBRA

Had a look ☐ Nearly there ☐ Nailed it! ☐

Real-life graphs

You can draw graphs to explain real-life situations. This graph shows the cost of buying some printed T-shirts from three different companies.

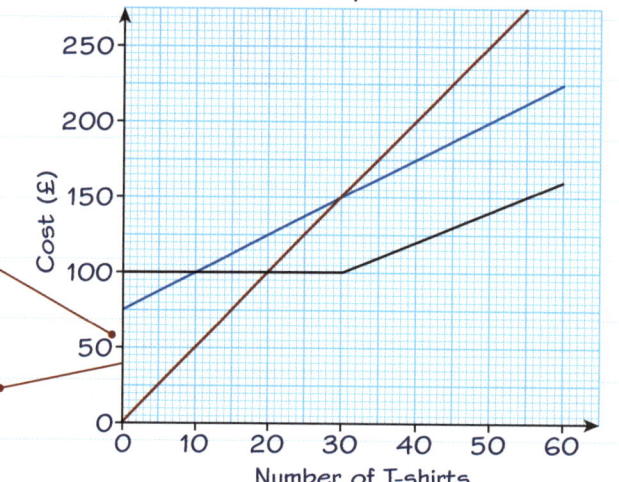

Be careful when reading scales on graphs. 10 small squares represent £50 so each small square represents £5.

Terry's T-shirts would be cheapest if you were ordering 10 T-shirts.

TERRY'S T-SHIRTS
No minimum order
£5 per shirt

SHIRT-O-GRAPH
Set-up cost £75
£2.50 per shirt

PAM'S PRINTING
Special offer 30 shirts for £100
Additional shirts just £2

Worked example *Target grade 3*

This graph shows the cost, C (£), of a taxi journey of length m miles.

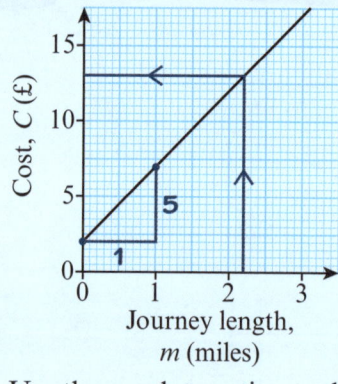

Everything in blue is part of the answer.

(a) Use the graph to estimate the cost of a journey of 2.2 miles. **(1 mark)**

£13

(b) The taxi company charges a fixed amount plus a cost per mile. Write down:
  (i) the fixed amount
  (ii) the cost per mile. **(2 marks)**

(i) £2   (ii) £5

Interpreting the gradient

You might need to **interpret the gradient** of a straight-line graph:

✓ Straight-line graphs have a **constant gradient**.

✓ The gradient tells you how much the **vertical** variable increases when you increase the **horizontal** variable by one unit.

✓ Pay close attention to the scale on the axes – don't just count squares on the graphs.

There's more about gradients on page 37.

The additional cost per mile is the **gradient** of the graph.

Now try this *Target grade 3*

A mobile phone company charges a fixed monthly price plus a cost per minute.
(a) What is the fixed monthly charge? **(1 mark)**
(b) What is the total cost if you use 400 minutes in one month? **(1 mark)**
(c) The phone company claims that 'For every 10 minutes you only pay an extra 25p.' Is the phone company's claim true? Give a reason for your answer. **(3 marks)**

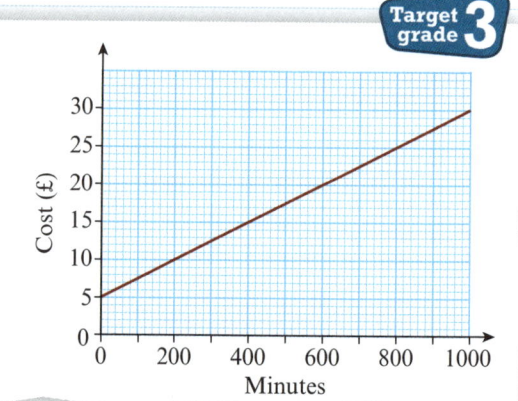

Had a look ☐   Nearly there ☐   Nailed it! ☐   ALGEBRA

Distance–time graphs

A **distance–time** graph shows how distance changes with time. This distance–time graph shows Jodi's run. The shape of the graph gives you information about the journey.

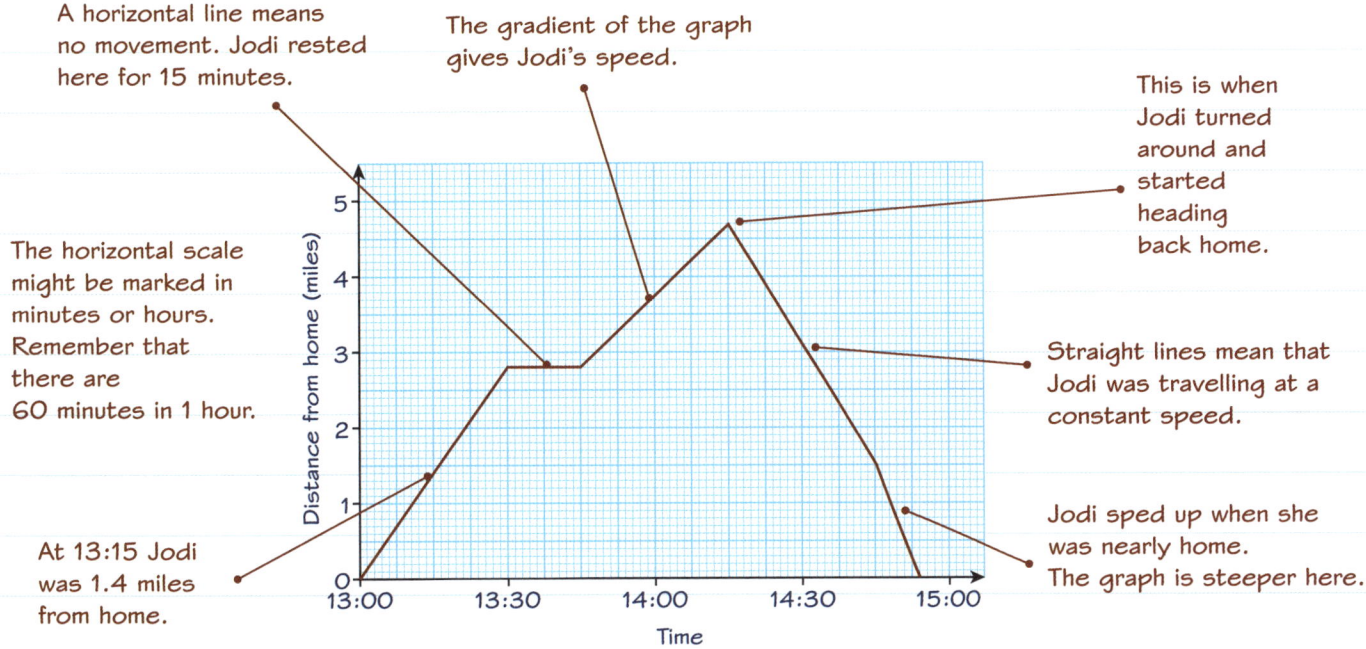

A horizontal line means no movement. Jodi rested here for 15 minutes.

The gradient of the graph gives Jodi's speed.

This is when Jodi turned around and started heading back home.

The horizontal scale might be marked in minutes or hours. Remember that there are 60 minutes in 1 hour.

Straight lines mean that Jodi was travelling at a constant speed.

At 13:15 Jodi was 1.4 miles from home.

Jodi sped up when she was nearly home. The graph is steeper here.

Worked example

This distance–time graph shows Nick's bike ride.
(a) For how long did Nick rest in total? **(1 mark)**

20 minutes

(b) During which section of his journey was Nick travelling fastest? Give a reason for your answer. **(2 marks)**

A. Because that is where the graph is steepest.

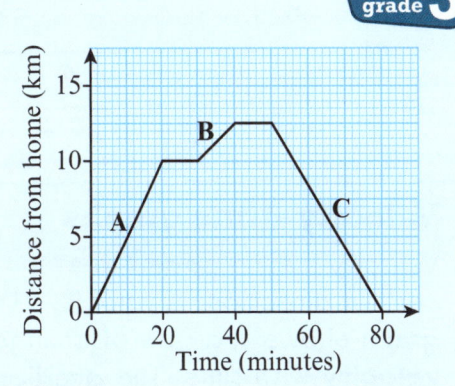

The flat sections of the graph are when Nick is resting. Read the scale on the graph carefully.

Now try this

Christina rode her bike to a friend's house. She rested once on the way, had coffee at her friend's house, then rode home.

(a) How long did Christina spend at her friend's house? **(1 mark)**
(b) Work out Christina's average speed for her return journey. **(2 marks)**

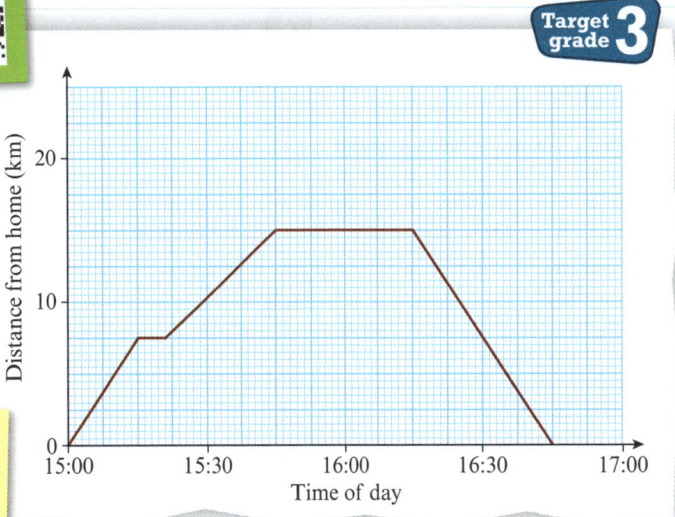

You can find the gradient of the graph, or you can use   Average speed = $\dfrac{\text{Distance}}{\text{Time}}$
There is more about speed on page 64.

Rates of change

A distance–time graph shows the **rate of change** of distance with time. You can revise distance–time graphs on page 41. You can use graphs to show other rates of change. These garden ponds are being filled with water. The graphs show the rate of change of the **depth** of the water in each pond.

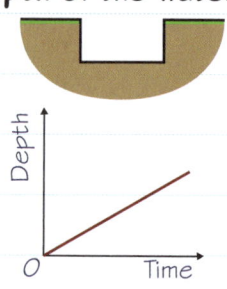 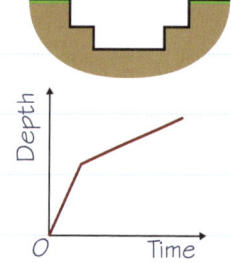 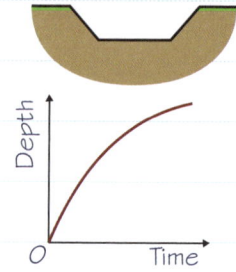

The narrower the section of pond, the quicker the water depth will increase.

Worked example

Supraj recorded the rate of rainfall, in mm per hour, during one afternoon.

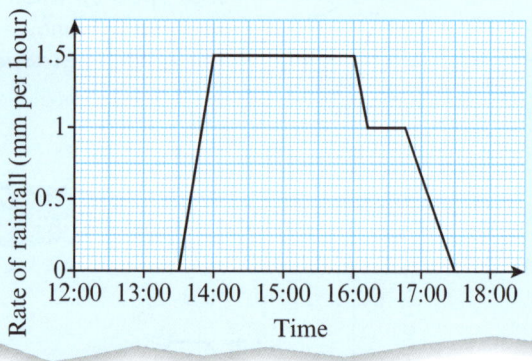

(a) At what time did it start to rain?
13:30

(b) Supraj said, 'It stopped raining at 14:00.' Is Supraj correct? Give a reason for your answer.
Supraj is wrong. The rain was falling steadily from 14:00 to 16:00. It didn't stop raining until 17:30.

This graph shows the **rate** of rainfall. The rain was getting harder from 13:30 to 14:00. It then fell at a steady rate until 16:00, when it started getting lighter.

Target grade **3**

Velocity–time graphs

Velocity means **speed** in a certain direction. A velocity–time graph shows the rate of change of **velocity** with time. The **gradient** on a velocity–time graph tells you the **acceleration**. You need to recognise these four situations:

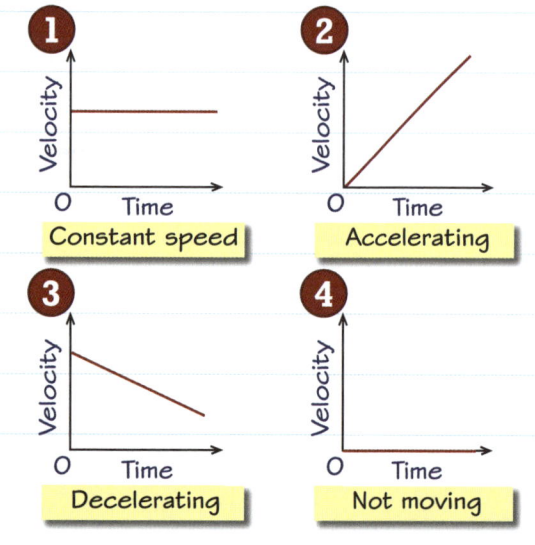

Now try this

This is the velocity–time graph for a car journey.

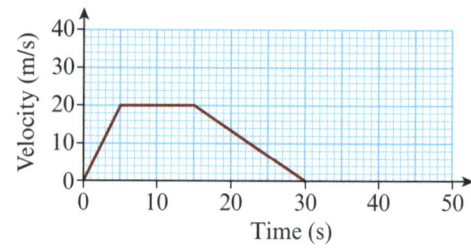

Which of these statements are true and which are false? Give reasons for your answers.

A The car travelled at a constant speed between 0 and 5 seconds.
B The car travelled at a constant speed between 5 and 15 seconds.
C The car was accelerating between 0 and 5 seconds.
D The car was not moving between 5 and 15 seconds.

(4 marks)

Target grade **4**

Had a look ☐ Nearly there ☐ Nailed it! ☐   **ALGEBRA**

Expanding double brackets

You need to be able to expand the **product** of two brackets. Use one of these three methods.

1 One-at-a-time
You can expand one bracket first, then the other:

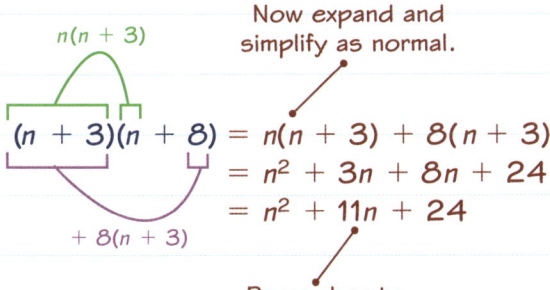

Now expand and simplify as normal.

$(n + 3)(n + 8) = n(n + 3) + 8(n + 3)$
$ = n^2 + 3n + 8n + 24$
$ = n^2 + 11n + 24$

Remember to collect like terms.

2 FOIL
The acronym **FOIL** tells you the four terms you need to find:

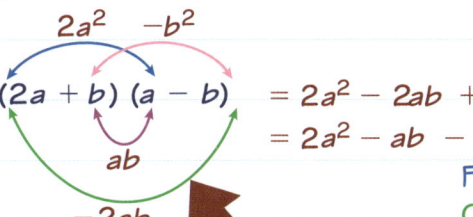

$(2a + b)(a - b) = 2a^2 - 2ab + ab - b^2$
$ = 2a^2 - ab - b^2$

First terms
Outer terms
Inner terms
Last terms

Some people remember this as a 'smiley face'.

3 Grid method
You can draw a grid to help you expand the brackets:

$(x + 7)(x - 5) = x^2 - 5x + 7x - 35$
$ = x^2 + 2x - 35$

Remember to collect like terms if possible.

	x	-5
x	x^2	$-5x$
7	$7x$	-35

The negative sign belongs to the 5. You need to write it in your grid.

Worked example Target grade 4

Expand and simplify $(y + 2)(y + 9)$ **(2 marks)**
$(y + 2)(y + 9) = y^2 + 9y + 2y + 18$
$ = y^2 + 11y + 18$

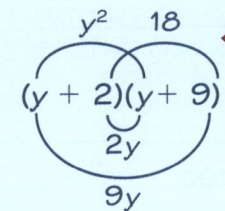

This example shows the **FOIL** method. The question says expand and **simplify** so collect like terms at the end.

Worked example Target grade 5

Expand and simplify $(3p - 4)^2$ **(2 marks)**

$(3p - 4)^2 = (3p - 4)(3p - 4)$
$ = 9p^2 - 12p - 12p + 16$
$ = 9p^2 - 24p + 16$

	$3p$	-4
$3p$	$9p^2$	$-12p$
-4	$-12p$	16

When you square a term in brackets you multiply the **whole bracket** by itself. Start by writing out:

$(3p - 4)^2 = (3p - 4)(3p - 4)$

This example shows the **grid method** but you could use any method to expand the brackets.

Remember that $3p \times 3p = 9p^2$ and be careful with the **negative signs**:
$-4 \times -4 = 16 \quad p \times -4 = -4p$

Now try this

1 Complete this working:
$(x + 6)(x + 3) = x^2 + 3x + \square x + \square$
$ = x^2 + \square x + \square$ **(2 marks)**

2 Expand and simplify
(a) $(x + 5)(x - 1)$ **(2 marks)**
(b) $(p - 6)^2$ **(2 marks)**

Target grade 4

Worked solution video

ALGEBRA

Had a look ☐   Nearly there ☐   Nailed it! ☐

Quadratic graphs

Quadratic equations contain an x^2 term. Quadratic equations have **curved** graphs. You can draw the graph of a quadratic equation by completing a table of values.

Worked example

The **turning point** is the point where the direction of the curve changes.

(a) Complete the table of values for $y = 4x - x^2$ **(2 marks)**

x	−1	0	1	2	3	4	5
y	−5	0	3	4	3	0	−5

(b) On the grid, draw the graph of $y = 4x - x^2$ **(2 marks)**
(c) Write down the coordinates of the turning point.
(2, 4)   **(1 mark)**

When $x = -1$: $4 \times (-1) - (-1)^2$
$= -4 - 1 = -5$

When $x = 4$: $4 \times 4 - 4^2 = 16 - 16 = 0$

Plot your points carefully on the graph and join them with a **smooth** curve.

Check it!
All the points on your graph should lie on the curve. If one of the points doesn't fit then double check your calculation.

Everything in blue is part of the answer.

Drawing quadratic curves
- ✓ Use a sharp pencil.
- ✓ Plot the points carefully.
- ✓ Draw a smooth curve that passes through every point.
- ✓ Label your graph.
- ✓ Shape of graph will be either ∪ or ∩

Drawing a smooth curve
It's easier to draw a smooth curve if you turn your graph paper so your hand is **inside** the curve.

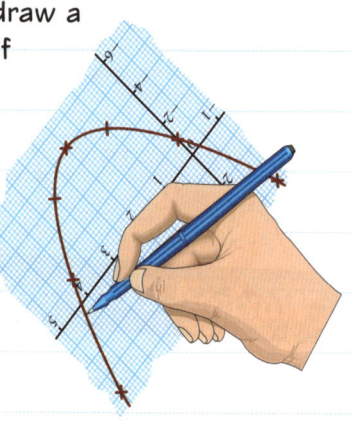

Now try this

Worked solution video

Target grade 4

(a) Complete this table of values for $y = x^2 + 2$ **(2 marks)**

x	−3	−2	−1	0	1	2	3
y		6		2	3	6	

(b) Draw the graph of $y = x^2 + 2$ for $x = -3$ to $x = 3$ **(2 marks)**

(c) Use your graph to find the value of y when $x = 2.5$ **(1 mark)**

Had a look ☐ Nearly there ☐ Nailed it! ☐   **ALGEBRA**

Using quadratic graphs

You might need to read values off a quadratic graph. You can do this by looking at the points where the graph crosses the *x*-axis, or by drawing a horizontal line on your graph.

Worked example
Target grade 4

(a) Complete the table of values for $y = x^2 - x - 3$
(2 marks)

x	−2	−1	0	1	2	3
y	3	−1	−3	−3	−1	3

(b) On the grid, draw the graph of $y = x^2 - x - 3$
(2 marks)

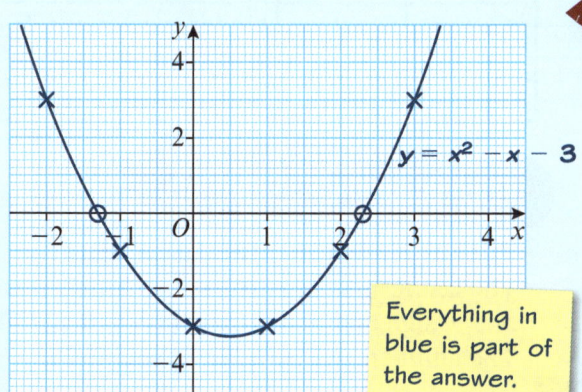

> This is an example of a **quadratic graph**. Its equation contains an x^2 term. You should always use a table of values to draw quadratic graphs.

Examiners' report

Don't use a ruler – you need to join your points with a **smooth curve**.

Real students have struggled with questions like this in recent exams – **be prepared!**

Everything in blue is part of the answer.

(c) Use your graph to write down the solutions of
$x^2 - x - 3 = 0$
(2 marks)

$x = -1.3$ and $x = 2.3$

> The solutions of the equation $x^2 - x - 3 = 0$ are the *x*-coordinates at the points where the graph crosses the *x*-axis. These are sometimes called the **roots** of the equation. You should always read graphs accurate to the nearest small square. So on this graph your answers will be accurate to 1 decimal place.

Now try this
Target grade 4

This is the graph of $y = x^2 - 4x - 1$
(a) Write down the values of *x* where the graph crosses the *x*-axis. **(2 marks)**
(b) Write down the values of *x* when $y = 8$ **(2 marks)**

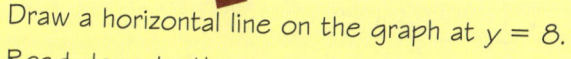

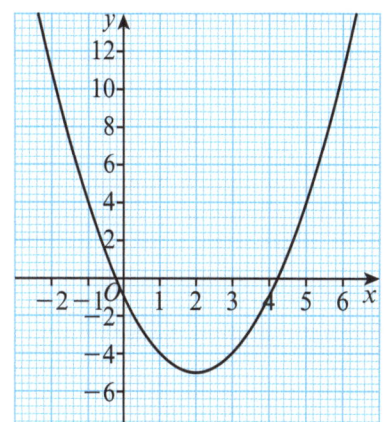

> Draw a horizontal line on the graph at $y = 8$.
> Read down to the *x*-axis at the points where the line crosses the curve.
> Read the graph correct to the nearest small square.
> These values of *x* are the solutions to the quadratic equation $x^2 - 4x - 1 = 8$

ALGEBRA   Had a look ☐   Nearly there ☐   Nailed it! ☐

Tricky Topic

Factorising quadratics

You need to be able to **factorise** quadratic expressions by writing them as the product of **two brackets**. Make sure you are really confident expanding **double brackets** before you start. You can revise this on page 43. You can also revise factorising simpler expressions on page 29.

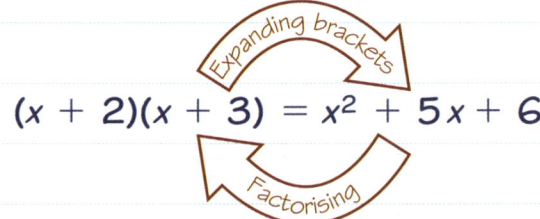

$(x + 2)(x + 3) = x^2 + 5x + 6$

Factorising $x^2 + bx + c$

You need to write the expression with **two brackets**. Look for two numbers whose **sum** is equal to b and whose **product** is equal to c.

You need to find two numbers which add up to 7... $5 + 2 = 7$

$x^2 + 7x + 10 = (x + 5)(x + 2)$

$5 \times 2 = 10$

... and multiply to make 10

You can use this table to help you find the two numbers.

b	c	Factors
positive	positive	both numbers positive
positive	negative	bigger number positive and smaller number negative
negative	negative	bigger number negative and smaller number positive
negative	positive	both numbers negative

Worked example — Target grade 5

Factorise $x^2 - x - 20$ **(2 marks)**

Factor pairs of 20:
1 and 20, 2 and 10, <u>4 and 5</u>

$x^2 - x - 20 = (x + 4)(x - 5)$

Check:
$(x + 4)(x - 5) = x^2 - 5x + 4x - 20$
$= x^2 - x - 20$ ✓

Examiners' report

The answer will have **two sets of brackets**. The last term is negative, so the brackets will have one + sign and one − sign: $(x + \square)(x - \square)$. The numbers will be a **factor pair of 20**.

With any factorisation, the safest thing to do is to **check your answer** by expanding the brackets.

Real students have struggled with questions like this in recent exams – **be prepared!**

Difference of two squares

You can factorise expressions that are written as
$(\text{something})^2 - (\text{something else})^2$

Use this rule:
$a^2 - b^2 = (a + b)(a - b)$
$x^2 - 36 = x^2 - 6^2$
$= (x + 6)(x - 6)$

36 is a square number.
$36 = 6^2$ so $a = x$ and $b = 6$

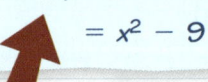

Worked example — Target grade 5

Factorise $x^2 - 9$ **(2 marks)**

$x^2 - 9 = (x + 3)(x - 3)$

Check:
$(x + 3)(x - 3) = x^2 - 3x + 3x - 9$
$= x^2 - 9$ ✓

Keep an eye out for the difference of two squares. $9 = 3^2$ so $a = x$ and $b = 3$.

Now try this — Target grade 5

1 Factorise
(a) $x^2 + 6x + 8$ **(2 marks)**
(b) $x^2 - 10x + 16$ **(2 marks)**

Worked solution video

2 Factorise
(a) $x^2 - 144$ **(2 marks)**
(b) $x^2 - 49$ **(2 marks)**

Had a look ☐   Nearly there ☐   Nailed it! ☐   **ALGEBRA**

Quadratic equations

Tricky Topic

You need to be able to **solve** a quadratic equation **without a calculator** by factorising one side. If you need to solve a quadratic equation in your Foundation GCSE exam it will look something like this:

The left-hand side will be a **quadratic expression**.

$$x^2 + 6x + 5 = 0$$

The right-hand side will be **zero**.

Follow the steps below to solve the equation.

1 Factorise the left-hand side of the equation:
$(x + 5)(x + 1) = 0$

2 Set each factor equal to zero:
$(x + 5)(x + 1) = 0$
$(x + 5) = 0 \quad (x + 1) = 0$

3 Solve the two linear equations to find the two solutions:
$x + 5 = 0 \qquad x + 1 = 0$
$x = -5 \qquad x = -1$

For a reminder about factorising quadratic expressions look at page 46.

Write both solutions in the form $x = \boxed{}$

Worked example *Target grade 5*

Solve $x^2 + 8x - 9 = 0$ **(3 marks)**

$(x + 9)(x - 1) = 0$

$x + 9 = 0 \qquad x - 1 = 0$
$x = -9 \qquad\quad x = 1$

1. Start by factorising the left-hand side of the equation. You can check by expanding the brackets:
$(x + 9)(x - 1) = x^2 - x + 9x - 9$
$\qquad\qquad\qquad = x^2 + 8x - 9$ ✓
2. Set each factor equal to zero.
3. Solve the linear equations to get **two** solutions.

Check it!
$(1)^2 + 8(1) - 9 = 1 + 8 - 9 = 0$ ✓
$(-9)^2 + 8(-9) - 9 = 81 - 72 - 9 = 0$ ✓

Worked example *Target grade 5*

Solve $x^2 - 100 = 0$ **(3 marks)**

$(x + 10)(x - 10) = 0$

$x + 10 = 0 \qquad x - 10 = 0$
$x = -10 \qquad\quad x = 10$

You can also use inverse operations:
$x^2 - 100 = 0 \qquad (+ 100)$
$x^2 = 100 \qquad\quad (\sqrt{})$
$x = 10 \text{ or } x = -10$
There are two answers because
$(-10)^2 = -10 \times -10 = 100$

Two to watch

1 When one solution is $x = 0$
$x^2 - 10x = 0$
$x(x - 10) = 0$
Solutions are $x = 0$ and $x = 10$

2 Difference of two squares
$x^2 - 25 = 0$
$(x - 5)(x + 5) = 0$
Solutions are $x = 5$ and $x = -5$

Now try this *Target grade 5*

1 (a) Factorise $x^2 + 5x + 6$ **(2 marks)**
  (b) Write down the two solutions of
     $x^2 + 5x + 6 = 0$ **(1 mark)**

Both solutions are **negative**.

2 Solve
  (a) $x^2 - 5x = 0$ **(2 marks)**
  (b) $x^2 + 3x - 28 = 0$ **(3 marks)**
  (c) $x^2 - 144 = 0$ **(3 marks)**

For parts (a) and (c) look at the blue box above.

47

ALGEBRA — Had a look ☐ Nearly there ☐ Nailed it! ☐

Cubic and reciprocal graphs

You might need to **draw** or **interpret** cubic and reciprocal graphs in your exam. You can use a **table of values** to draw any graph, but it helps if you know what the general shape of the graph is going to be.

1 Cubic graphs

Graphs that contain an x^3 term and no higher powers of x are called cubic graphs. Here are two examples:

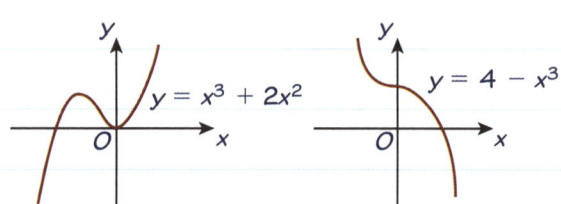

2 The reciprocal graph

The graph of $y = \dfrac{1}{x}$ is called the reciprocal graph.

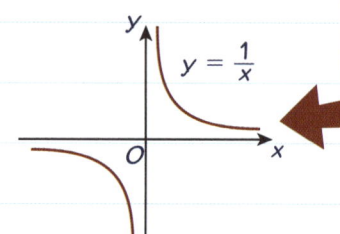

The graph gets closer and closer to the x- and y-axes but never touches them.

Worked example

(a) Complete the table of values for $y = x^3 - 4x - 3$

x	−2	−1	0	1	2	3
y	−3	0	−3	−6	−3	12

(2 marks)

(b) On the grid, draw the graph of $y = x^3 - 4x - 3$ for $-2 \leq x \leq 3$ **(2 marks)**

(c) (i) Estimate the value of x when $y = 6$ **(1 mark)**

2.7

(ii) Comment on the accuracy of your estimate. **(1 mark)**

The estimate is not very accurate because it is based on reading off a graph.

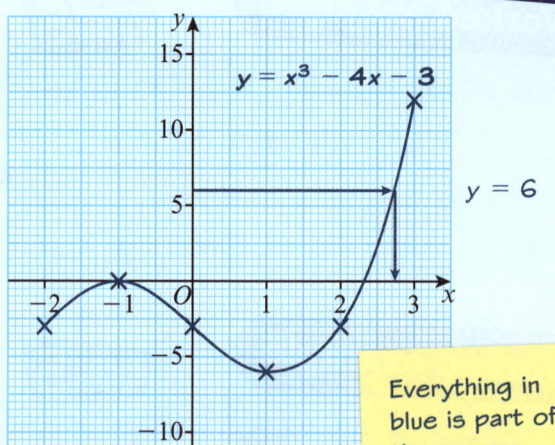

Everything in blue is part of the answer.

This is a **cubic graph** with a **positive** coefficient of x. If you recognise the shape of the graph then it's easier to tell if you've plotted your coordinates correctly.

Now try this

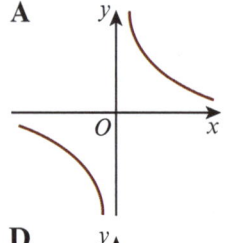

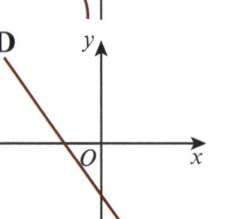

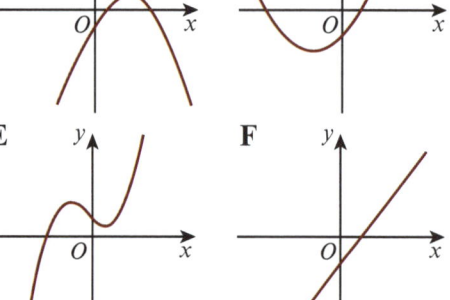

Write down the letter of the graph which could have the equation

(a) $y = -4x - 2$

(b) $y = x^3 - 3x + 4$

(c) $y = \dfrac{1}{x}$

(d) $y = 8x - 15 - x^2$

(e) $y = 2x - 3$

(f) $y = x^2 + 2x - 8$ **(6 marks)**

Had a look ☐   Nearly there ☐   Nailed it! ☐   **ALGEBRA**

Simultaneous equations

Simultaneous equations have two unknowns. You need to find the values for the two unknowns which make **both** equations true.

Algebraic solution

1. Number each equation.
2. If necessary, multiply the equations so that the coefficients of one unknown are the same.
3. Add or subtract the equations to **eliminate** that unknown.
4. Once one unknown is found use substitution to find the other.
5. Check the answer by substituting both values into the original equations.

$3x + y = 20$ ①
$x + 4y = 14$ ②
$12x + 4y = 80$ ① × 4
$- (x + 4y = 14)$ $-$②
$11x = 66$
$x = 6$

Substitute $x = 6$ into ①:
$3(6) + y = 20$
$18 + y = 20$
$y = 2$
Solution is $x = 6$, $y = 2$.
Check: $x + 4y = 6 + 4(2) = 14$ ✓

Worked example Target grade 5

Solve the simultaneous equations
$6x + 2y = -3$ ①
$4x - 3y = 11$ ② **(4 marks)**

$18x + 6y = -9$ ① × 3
$+ 8x - 6y = 22$ ② × 2
$26x = 13$
$x = \frac{1}{2}$

Substitute $x = \frac{1}{2}$ into ①:
$6(\frac{1}{2}) + 2y = -3$
$3 + 2y = -3$
$2y = -6$
$y = -3$
Solution is $x = \frac{1}{2}$, $y = -3$

Easier eliminations

You can save time by choosing the right unknown to eliminate. Look for one of these:

① If an unknown appears **on its own** in one equation you only need to multiply one equation to eliminate that unknown.

② If an unknown has **different signs** in the two equations you can eliminate by **adding**.

Multiply both equations by a whole number to make the coefficients of y the same.

Check it!
Always use the equation you **didn't** substitute into to check your answer:
$4x - 3y = 4(\frac{1}{2}) - 3(-3) = 2 + 9 = 11$ ✓

Graphical solution

You can solve these simultaneous equations by drawing a graph.
$x - y = 1$ $x + 2y = 4$
The coordinates of the point of intersection give the solution to the simultaneous equations.
The solution is $x = 2$, $y = 1$

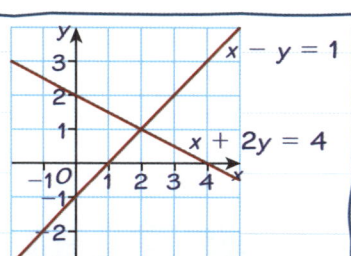

Now try this Target grade 5

1. Solve the simultaneous equations
   $3x - 2y = 12$
   $x + 4y = 11$ **(3 marks)**

2. By drawing two suitable straight lines on a coordinate grid, solve these simultaneous equations.
   $3y + 2x = 6$
   $y = 2x - 2$ **(4 marks)**

Draw a coordinate grid from -3 to 5 in both directions.

Worked solution video

ALGEBRA

Had a look ☐   Nearly there ☐   Nailed it! ☐

Tricky Topic

Rearranging formulae

Most formulae have one letter on its own on one side of the formula. This letter is called the **subject** of the formula.

$P = 2l + 2w$ P is the subject.
$A = \frac{1}{2}bh$ A is the subject.

You can use algebra to change the subject of a formula. This is like solving an equation.

> If you need to find a missing value which is not the subject in a formula:
> 1. Substitute any values you know into the formula.
>   For a reminder about substitution have a look at page 25.
> 2. Solve the equation to find the missing value.

Worked example — Target grade 3

Alicia works h hours of normal time and v hours of overtime each week. She is paid £P. She uses this formula to work out her pay:
$P = 8h + 12v$
Last week Alicia worked 32 hours of normal time and was paid £328. How many hours of overtime did she work? **(3 marks)**

$328 = 8 \times 32 + 12v$
$328 = 256 + 12v$ (-256)
$72 = 12v$ $(\div 12)$
$6 = v$

Alicia worked 6 hours of overtime.

To make p the subject of this formula you have to do the same thing to **both sides** of the formula until you have p on its own on one side.

$N = 2p + 3q^2$ $(-3q^2)$ — Subtract any terms you don't need.

You have to divide **everything** on this side by 2.

$N - 3q^2 = 2p$ $(\div 2)$ — $2p$ means $2 \times p$ so divide both sides by 2 to get p on its own.

$\frac{N - 3q^2}{2} = p$

Worked example — Target grade 3

Rearrange this formula to make Q the subject:
$R = 1 - 5Q$ **(2 marks)**

$R = 1 - 5Q$ $(+5Q)$
$R + 5Q = 1$ $(-R)$
$5Q = 1 - R$ $(\div 5)$
$Q = \frac{1 - R}{5}$

> You want the term containing Q to be positive. Start by adding $5Q$ to both sides.

Worked example — Target grade 3

Rearrange this formula to make h the subject:
$A = \frac{1}{2}bh$ **(2 marks)**

$A = \frac{1}{2}bh$ $(\times 2)$
$2A = bh$ $(\div b)$
$\frac{2A}{b} = h$

> You need the new subject on its own on one side of the formula. It doesn't matter which side.

Now try this

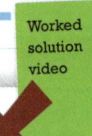

Worked solution video

Target grade 3

1 (a) Rearrange this formula to make Q the subject:
$W = 7Q - 100$ **(2 marks)**

Target grade 4

(b) Rearrange this formula to make V the subject:
$d = \frac{m}{V}$ **(2 marks)**

2 Rearrange this formula to make n the subject:
$p = 2(m - n)$ **(2 marks)**

> Start by expanding the brackets.
> For a reminder about expanding brackets look at page 28.

50

Had a look ☐   Nearly there ☐   Nailed it! ☐   **ALGEBRA**

Using algebra

You can sometimes use the information given in a question to write an equation. You might need to choose a letter to represent an unknown quantity you are trying to find.

Worked example   Target grade 3

Here are three number cards.

Each card has a whole number written on the back.
The number on card **B** is twice the number on card **A**.
The number on card **C** is five more than the number on card **A**.
The sum of the numbers on all three cards is 37.
Work out the number on each card. **(4 marks)**

$B = 2A$ $C = A + 5$
$A + B + C = 37$
$A + 2A + (A + 5) = 37$
$4A + 5 = 37$ $(- 5)$
$4A = 32$ $(\div 4)$
$A = 8$
$B = 2 \times 8 = 16$ $C = 8 + 5 = 13$

Problem solved! You could try guessing the numbers on the cards, but it might take a long time. You can solve this problem quickly by writing your own equation.

- Use A to represent the number on card **A**.
- Write expressions for the numbers on cards **B** and **C** in terms of A.
- Add together these three expressions and set the total equal to 37. Solve this equation to find A.

Once you have calculated A, remember to work out the values of B and C as well.

Check it!
$8 + 16 + 13 = 37$ ✓

True or false?

You might need to **substitute** values into an algebraic statement to show that it is **false**. You only need to find one **counter-example** to show that a statement is false.

Try some different values of n and write down all your working.
You should know the prime numbers up to 20.

Worked example   Target grade 3

n is a whole number.
Supraj says that $2n^2 + 3$ is always a prime number.
Give an example to show that he is wrong. **(2 marks)**

$n = 1$: $2n^2 + 3 = 2 \times 1 + 3 = 5$ Prime
$n = 2$: $2n^2 + 3 = 2 \times 4 + 3 = 11$ Prime
$n = 3$: $2n^2 + 3 = 2 \times 9 + 3 = 21 = 3 \times 7$
   Not prime

Now try this

1 These two rectangles have the same perimeter.

In the diagram, all of the measurements are in metres. Work out the width and height of rectangle **A**. Show all of your working. **(4 marks)**

2 Jill says that the expression $n^2 + 4n + 3$, where n is an integer, is never a multiple of 7. Give an example to show that Jill is wrong. **(2 marks)**

Start by writing expressions for the perimeter of each rectangle. The perimeter of rectangle A is:
$(x - 3) + 2x + (x - 3) + 2x$
Remember to answer the question at the end.

ALGEBRA

Had a look ☐   Nearly there ☐   Nailed it! ☐

Identities and proof

An **identity** is something that is **always** true. The right-hand side will always be exactly equal to the left-hand side, no matter what values you substitute. You use the symbol '≡' to represent an identity. You might need to **show** that an identity is true in your exam.

Golden rule
An identity is not like an equation. Do not solve it using the balance method.
Manipulate each side separately. ✓
Apply the same operation to both sides. ✗

Worked example (Target grade 5)

Show that $(n - 1)^2 + (n + 1)^2 \equiv 2(n^2 + 1)$
(2 marks)

$(n - 1)^2 + (n + 1)^2$
$\equiv (n - 1)(n - 1) + (n + 1)(n + 1)$
$\equiv n^2 - 2n + 1 + n^2 + 2n + 1$
$\equiv 2n^2 + 2$
$\equiv 2(n^2 + 1)$

If you need to **show** that an identity is true in an exam you should show every line of your working.

Start with the expression on the left-hand side. Use multiplying out, simplifying and factorising to work towards the expression on the right-hand side.

Proving results with algebra

You can use algebra to show that numbers are **odd**, **even**, or **multiples** of another number. You can use n to represent any integer. You might need to **rearrange** or **simplify** algebraic expressions, and you might have to write a **conclusion**. Use the table on the right to help you.

Number fact	Written using algebra
even number	$2n$
odd number	$2n + 1$ or $2n - 1$
multiple of 3	$3n$
consecutive numbers	$n, n + 1, n + 2, \ldots$
consecutive even numbers	$2n, 2n + 2, 2n + 4, \ldots$
consecutive odd numbers	$2n + 1, 2n + 3, 2n + 5, \ldots$
consecutive square numbers	$n^2, (n + 1)^2, (n + 2)^2, \ldots$

Multiple or not?

If n is an integer then $3n$ is a **multiple of 3**. This means that $3n + 1$ and $3n + 2$ are **not** multiples of 3.

Problem solved! You can show that this is true for **any** three consecutive integers using algebra.
1. Write the first integer as n and the next two integers as $n + 1$ and $n + 2$.
2. Write an expression for the sum of your three integers. Simplify your expression.
3. Factorise your expression then write a conclusion.

Worked example (Target grade 5)

Show that the sum of any three consecutive integers is always a multiple of 3. **(3 marks)**

$n, n + 1$ and $n + 2$ represent any three consecutive integers.
$n + (n + 1) + (n + 2) = 3n + 3$
$ = 3(n + 1)$
$n + 1$ is an integer, so $3(n + 1)$ is a multiple of 3.

Now try this (Target grade 5)

1 Show that $(a + b)(a - b) \equiv a^2 - b^2$
(2 marks)

2 Given that $4(x - n) = 3x - 1$ where n is an integer, show that x is an odd number. **(3 marks)**

Had a look ☐   Nearly there ☐   Nailed it! ☐   ALGEBRA

Problem-solving practice 1

About half of the questions in your Foundation GCSE exam will require you to **problem-solve, reason, interpret** or **communicate** mathematically. If you come across a tricky or unfamiliar question in your exam you can try some of these strategies:

- ✓ Sketch a diagram to see what is going on.
- ✓ Try the problem with smaller or easier numbers.
- ✓ Plan your strategy before you start.
- ✓ Write down any formulae you might be able to use.
- ✓ Use x or n to represent an unknown value.

AO2

AO3

Now try this

1 In the diagram, $ABCD$ is a rectangle.

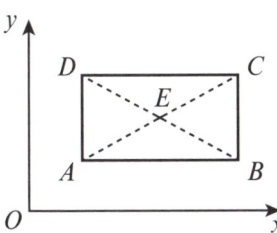

The length AB is twice the length AD.
E is the mid-point of the diagonals of the rectangle.
The coordinates of E are $(30, 20)$.
Work out one set of possible coordinates for points A, B, C and D. **(4 marks)**

Coordinates page 36   Target grade 3

There are lots of different correct answers to this question. Start by picking coordinates for A. In order for the width of the rectangle to be twice its height, the horizontal distance from A to E must be twice the vertical distance from A to E.

TOP TIP

Choose simple numbers to make your calculations easier. The coordinates of the mid-point are multiples of 10, so try using multiples of 10 for the coordinates of the vertices.

2 Dexter uses this formula to work out how long to cook a chicken.

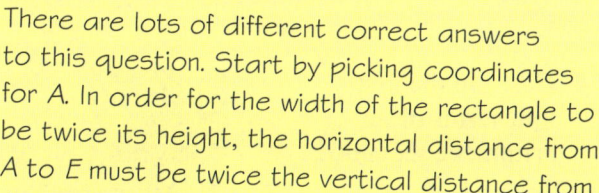

He works out that his chicken will take exactly 1 hour 33 minutes to cook.
Work out the weight of Dexter's chicken. **(3 marks)**

Rearranging formulae page 50   Target grade 3

The units in the formula are minutes, so write 1 hour 33 minutes as a time in minutes. To find the weight of Dexter's chicken, substitute any values you know into the formula, and then solve the equation to find the missing value.

TOP TIP

You can represent the chicken's weight in kg as a single letter, for example w, in order to simplify your equation.

Problem-solving practice 2

Now try this

3

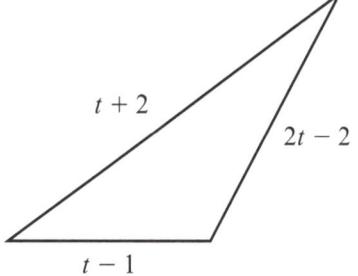

The perimeter of this triangle is 13 cm.
All lengths on the diagram are in cm.
Work out the value of t. **(4 marks)**

Linear equations 2 page 31
Using algebra page 51
Target grade 3

Use the information in the question to write an equation. Solve your equation to work out the value of t.

TOP TIP

Remember to simplify your equation by collecting like terms before solving.

---

4 Here are the first four terms of an arithmetic sequence:
9, 15, 21, 27, …
Explain why the number 65 cannot be a term of this sequence. **(3 marks)**

Worked solution video

Sequences 2 page 35
Target grade 4

Work out the nth term of the sequence. Then work out consecutive terms of the sequence on either side of 65.

TOP TIP

You can use any successful strategy to answer a question, as long as you show your method. You could set the nth term equal to 65 and solve an equation to show that n is not an integer.

---

5 A train travels a distance of y km in 3 minutes.
Show that the average speed of the train is $20y$ km/h. **(3 marks)**

Formulae page 26   Speed page 64
Target grade 4

You could convert minutes to hours then use the formula for average speed. Or you could find an expression for the distance travelled in 1 minute then multiply by 60.

TOP TIP

Try changing y into a number and calculating the average speed. It will give you an idea of how the formula works.

---

6 A straight line has a gradient of 8 and passes through the point (5, 10).
(a) Determine whether the point (10, 50) lies on the line.
You must show your working. **(3 marks)**
(b) Is your answer to part (a) reliable? Explain your answer. **(1 mark)**

Straight-line graphs 2 page 39
Target grade 5

You could tackle this question using algebra, or by drawing a diagram to scale. Whichever method you use, make sure you write a conclusion.

TOP TIP

If you have to comment on **reliability** then think about the method you used. If you used a diagram your answer might be less reliable than if you used algebra.

Had a look ☐ Nearly there ☐ Nailed it! ☐

RATIO & PROPORTION

Percentages

'**Per cent**' means '**out of 100**'. You can write a percentage as a fraction over 100.

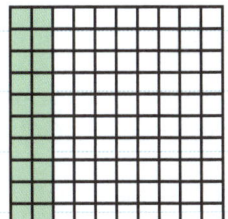 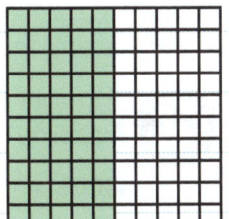  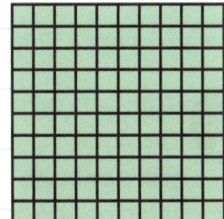

$20\% = \frac{20}{100} = \frac{1}{5}$ $50\% = \frac{50}{100} = \frac{1}{2}$ $75\% = \frac{75}{100} = \frac{3}{4}$ $100\% = \frac{100}{100} = 1$

Percentages with a calculator

To find a percentage of an amount:

| Divide the percentage by 100 |
↓
| Multiply by the amount |

For example, 12% of 80 cm is 9.6 cm.
12 ÷ 100 = 0.12
0.12 × 80 = 9.6

First work out 15% of £120. Then subtract this from £120.

You can also find 15% of £120 by working out 1% and then multiplying by 15.

Worked example (Target grade 3)

A car rental company reduces its prices by 15% in a sale.
A car normally costs £120 per week to rent. Work out the weekly rental cost of a car in the sale. **(3 marks)**

15 ÷ 100 = 0.15
0.15 × 120 = 18
120 − 18 = 102
The car costs £102 per week in the sale.

To write one quantity as a percentage of another:

| Divide the first quantity by the second quantity |
↓
| Multiply your answer by 100 |

For example, 3 out of 12 yoghurts in a pack are strawberry.
3 ÷ 12 = 0.25
0.25 × 100 = 25
So 25% of the yoghurts are strawberry.

Worked example (Target grade 3)

In a year group of 96 students, 60 own a bicycle.
Express 60 as a percentage of 96. **(2 marks)**

60 ÷ 96 = 0.625
0.625 × 100 = 62.5
62.5% of the students own a bicycle.

Now try this

1% of £18 200 is £182, so 3% is 3 lots of £182.

Target grade 1
1 Sam earns £18 200 a year. He is given a pay rise of 3%. How much is his pay rise? **(2 marks)**

Target grade 3
2 A family recycle 8 kg of waste in a week. 2.8 kg of this waste is paper. What percentage of the recycled waste is paper? **(2 marks)**

 **RATIO & PROPORTION**

Had a look ☐ Nearly there ☐ Nailed it! ☐

Fractions, decimals and percentages

Here are three important facts about fractions, decimals and percentages:

1 You can convert a decimal to a percentage by multiplying by 100.

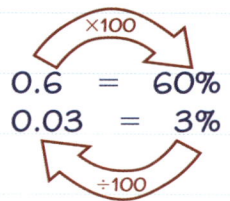

0.6 = 60%
0.03 = 3%

2 You can write any percentage as a fraction with denominator 100.

$60\% = \frac{60}{100} = \frac{6}{10} = \frac{3}{5}$

Simplify your fraction as much as possible.

3 Remember these common fraction, decimal and percentage equivalents.

Fraction	$\frac{1}{100}$	$\frac{1}{10}$	$\frac{1}{5}$	$\frac{1}{4}$	$\frac{1}{2}$	$\frac{3}{4}$
Decimal	0.01	0.1	0.2	0.25	0.5	0.75
Percentage	1%	10%	20%	25%	50%	75%

You can arrange a list of fractions, decimals and percentages in order of size by changing them to the same type.

Worked example *Target grade 3*

Carla has a bag of jelly beans.
15% of the jelly beans are strawberry.
$\frac{1}{4}$ of the jelly beans are pineapple.
$\frac{2}{5}$ of the jelly beans are apple.
The remaining jelly beans are cinnamon.
What percentage of the jelly beans are cinnamon? **(3 marks)**

$\frac{1}{4} = 25\%$

$\frac{2}{5} = 40\%$

15 + 25 + 40 = 80

100 − 80 = 20

20% of the jelly beans are cinnamon.

Problem solved! You might need to combine fractions, decimals and percentages in a word problem like this. Read the **whole question** before doing any working. The answer needs to be a **percentage** so the quickest way of answering this question would be to convert both fractions into percentages.

Remember that all of the jelly beans is 100%, so work out the sum of the other percentages then subtract that from 100% to find the percentage of jelly beans that are cinnamon.

Now try this

 *Target grade 2*

1 Write these percentages as fractions in their simplest form:
(a) 15% **(1 mark)**  $15\% = \frac{15}{100} = \frac{......}{20}$
(b) 68% **(1 mark)**

2 Write these numbers in order, starting with the smallest:
0.42 $\frac{2}{5}$ 36%   **(2 marks)**

 Write all three numbers as decimals then compare.

 *Target grade 3*

3 In a game, 5 players each put 12 counters on a table.
Sasha wins 35% of the counters.
Peter wins $\frac{2}{5}$ of the counters.
Haydon wins the remaining counters.
How many counters does Haydon win?
Show all of your working. **(3 marks)**

 Worked solution video

Had a look ☐   Nearly there ☐   Nailed it! ☐

RATIO & PROPORTION

Percentage change 1

Here are two methods that you can use to increase or decrease an amount by a percentage.

Method 1
Work out 26% of £280:
$\frac{26}{100} \times £280 = £72.80$
Subtract the decrease:
£280 − £72.80 = £207.20

Method 2
Use a multiplier:
100% + 30% = 130%
$\frac{130}{100} = 1.3$
So the multiplier for a 30% increase is 1.3:
400 g × 1.3 = 520 g

Worked example — Target grade 3

A football club increases the prices of its season tickets by 5.2% each year.

In 2011 a top-price season ticket cost £650. Calculate the price of this season ticket in 2012.   **(2 marks)**

$\frac{5.2}{100} \times £650 = £33.80$

£650 + £33.80 = £683.80

When working with money, you must give your answer to 2 decimal places.

Check it!
Use common sense to check your answer. 5.2% is a small increase, so the answer should be a bit more than £650. ✓

Examiners' report

This question would appear on a calculator paper. Don't try to use a mental method or a written method – use your calculator!

Real students have struggled with questions like this in recent exams – **be prepared!**

Calculating a percentage increase or decrease

Work out the amount of the increase or decrease
↓
Write this as a percentage of the original amount

 Was £60 Now £39

60 − 39 = 21
$\frac{21}{60} = 35\%$

This is a 35% decrease.

A question might ask you to calculate a percentage **profit** or **loss** rather than an increase or decrease.

For a reminder about writing one quantity as a percentage of another, have a look at page 55.

Now try this

 Worked solution video

Reduction means decrease. Work out the decrease as a percentage of the original price.

1 (Target grade 3) In 2010, a business used 1755 reams of paper. In 2011 the same business reduced its paper usage by 23%. Work out how many reams of paper the business used in 2011. Give your answer to the nearest whole number.   **(2 marks)**

2 (Target grade 4) A TV originally cost £520. In a sale it was priced at £340. What was the percentage reduction in the price? Give your answer to 1 decimal place.   **(3 marks)**

RATIO & PROPORTION Had a look ☐   Nearly there ☐   Nailed it! ☐

Percentage change 2

You need to be able to calculate a percentage increase or decrease **without a calculator**.

Calculating percentages

You can use multiples of 1% and 10% to calculate percentages without a calculator.

Work out 12.5% of £600

10% of £600 is £60   600 ÷ 10 = 60
1% of £600 is £6   600 ÷ 100 = 6
0.5% of £600 is £3   6 ÷ 2 = 3

So, 12.5% of £600 is
£60 + £6 + £6 + £3 = £75

```
           Work out
        the percentage
    DECREASE        INCREASE
Subtract it from the   Add it to the
 original amount      original amount
```

Worked example *Target grade 3*

Which television is cheaper?

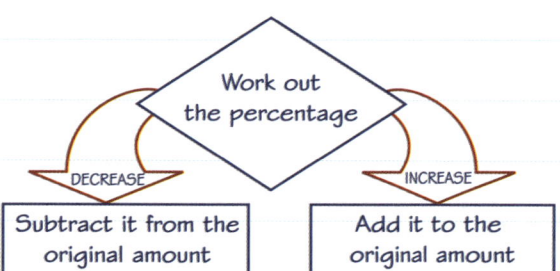

A £440 +20% VAT
B £550 SALE 2.5% OFF

Give reasons for your answer. **(5 marks)**

Television A
10% = £44
20% = £88
£440 + £88 = £528

Television B
1% = £5.50
0.5% = £2.75
£550 − £5.50 − £5.50 − £2.75
= £536.25

Television A is cheaper.

Examiners' report

The best way to give a reason for your answer is to show **all your working**. Make sure you lay out your working neatly and use the correct units.

Television A
1. Divide by 10 to calculate 10% of £440.
2. Multiply by 2 to work out 20%.
3. Add this to the original price.

Television B
1. Divide £550 by 100 to calculate 1%.
2. Divide by 2 to work out 0.5%.
3. Subtract two lots of £5.50 and one lot of £2.75 from the original price.

Don't just circle the cheaper television. You need to write your conclusion **in words**.

Real students have struggled with questions like this in recent exams – **be prepared!**

Percentage increase and decrease questions come in lots of different forms.

PERCENTAGE INCREASE   **PERCENTAGE DECREASE**

INTEREST   PAY RISE   SALES   DEPRECIATION

FIXED RATE ISA 2.8%

  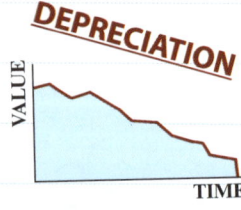

Now try this *Target grade 3*

Simon sees the same model of digital camera for sale in two different shops. Which shop is selling the camera at the cheaper price?
Give reasons for your answer. **(5 marks)**

CRUKS CAMERAS 30% OFF NORMAL PRICE OF £245

Spivs Cameras 35% OFF NORMAL PRICE OF £270

Had a look ☐   Nearly there ☐   Nailed it! ☐

RATIO & PROPORTION

Ratio 1

Ratios are used to compare quantities.

The ratio of apples to oranges is $3:2$
The ratio $3:2$ is in its simplest form.

You can write a ratio as a fraction.
$\frac{2}{5}$ of the pieces of fruit are oranges.

$3 + 2 = 5$
The denominator is the sum of the parts in the ratio.

Equivalent ratios

You can find equivalent ratios by multiplying or dividing by the same number.

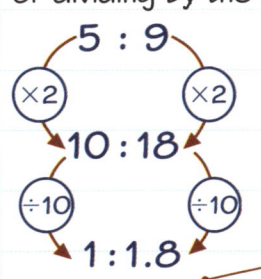

This equivalent ratio is in the form $1:n$
This is useful for calculations.

Worked example   Target grade 3

Jess and Simon are buying a new computer.
They pay in the ratio $2:3$.
Jess pays £194.
How much does Simon pay?   **(3 marks)**

$194 \div 2 = 97$
$3 \times 97 = 291$

Simon pays £291.

The order the people are written in is the same as the order of the numbers in the ratio. So 2 parts of the ratio represent the amount Jess paid. Divide £194 by 2 to work out how much each part of the ratio is worth, then multiply this value by 3 to find out how much Simon pays.

You can use a calculator, but remember to write down all your working.

Worked example   Target grade 2

A school has 200 students. 120 are female. Work out the ratio of male to female students. Give your answer in its simplest form.   **(2 marks)**

$200 - 120 = 80$

male : female
$80 : 120$
$\div 40 \qquad \div 40$
$2 : 3$

Simplest form

To write a ratio in its simplest form, find an equivalent ratio with the smallest possible whole number values.

Simplest form
$5:1 \quad 10:9$
$2:3:4$

Not simplest form
$1:1.5:2 \quad 10:2$
$1:0.9$

Now try this

Worked solution video

1 Target grade 3

(a) A bag contains white and pink marshmallows in the ratio $3:4$. What fraction of the marshmallows are pink? **(1 mark)**

(b) In a second bag, $\frac{2}{5}$ of the marshmallows are white and the rest are pink. What is the ratio of white to pink marshmallows in the second bag? **(1 mark)**

2 Target grade 2

Write the ratio $84:120$ in its simplest form. **(2 marks)**

3 Target grade 4

A bag contains red counters and green counters. The ratio of red to green counters is $4:3$. There are 32 red counters in the bag. What is the total number of counters in the bag? **(3 marks)**

RATIO & PROPORTION

Had a look ☐ Nearly there ☐ Nailed it! ☐

Ratio 2

Ratio is used in lots of problem-solving questions. You need to be able to answer ratio questions like the ones on this page **without using a calculator**. Remember that you won't have one for Paper 1.

Golden rule
You can answer lots of ratio questions by working out what **one part** of the ratio represents.

Worked example

Alexis, Nisha and Paul share a flat. One month their phone bill is £120.
They decide to split the bill in the ratio 3 : 5 : 2
How much does each person pay? **(3 marks)**

3 + 5 + 2 = 10
120 ÷ 10 = 12
3 × 12 = 36. Alexis pays £36.
5 × 12 = 60. Nisha pays £60.
2 × 12 = 24. Paul pays £24.

To divide a quantity in a given ratio:
1. Work out the total number of parts in the ratio.
2. Divide the quantity by this total.
3. Multiply your answer by each part of the ratio.

The order the people are written in is the same as the order of the numbers in the ratio. Alexis is first in the list, so 3 parts of the ratio represent the amount she pays.

Check it!
£36 + £60 + £24 = £120 ✓

Worked example

Jamie and Chaaya took part in a sponsored swim to raise money for charity.
The ratio of Jamie's total to Chaaya's total is 5 : 7
Chaaya raised £12 more than Jamie.
How much money did they raise in total?
(3 marks)

7 − 5 = 2
2 parts = £12 so 1 part = £6
Jamie = £30
Chaaya = £42
Total = £30 + £42 = £72

Problem solved! Start by working out what one part of the ratio represents. Chaaya raised £12 more than Jamie so **two parts** of the ratio represent £12. Therefore, **one part** of the ratio represents £6.
So Jamie raised 5 × £6 = £30
Chaaya raised 7 × £6 = £42

Check it!
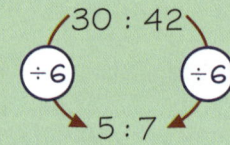

£42 − £30 = £12 ✓

Now try this

1. Ruth, Sue and Tess share £400 between them.
   Ruth receives £80 more than Sue.
   The ratio of Ruth's share to Sue's share is 9 : 5
   Work out how much Tess receives.
   (3 marks)

2. Terri mixed 300 g of rice with 240 g of fish.
   She added some onion to the mixture.
   The ratio of the weight of fish to the weight of onion was 3 : 2
   Work out the ratio of the weight of rice to the weight of onion. **(3 marks)**

Worked solution video

Had a look ☐ Nearly there ☐ Nailed it! ☐

RATIO & PROPORTION

Metric units

You can convert between metric units by multiplying or dividing by 10, 100 or 1000.

Length
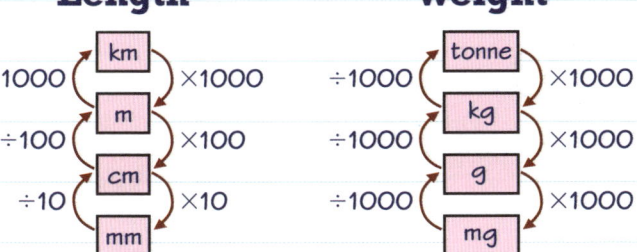

Weight

Volume or capacity
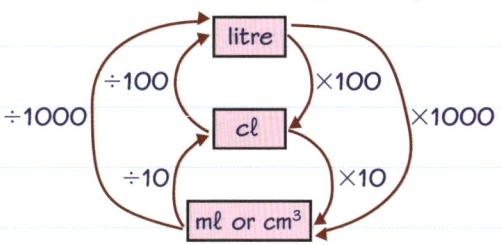

Place value diagrams

You can use a place value diagram to help you multiply and divide by 10, 100 and 1000.

For more on place value look at page 1.

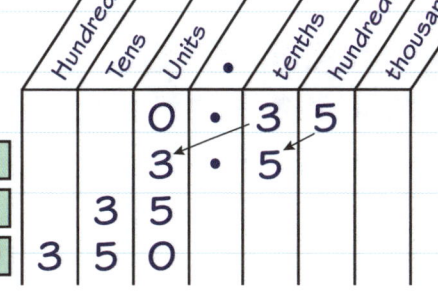

To multiply by 10, 100 or 1000 you move the digits 1, 2 or 3 places to the left.

$0.35 \times 10 = 3.5$
$0.35 \times 100 = 35$
$0.35 \times 1000 = 350$

To divide by 10, 100 or 1000 you move the digits 1, 2 or 3 places to the right.

$120 \div 10 = 12$
$120 \div 100 = 1.2$
$120 \div 1000 = 0.12$

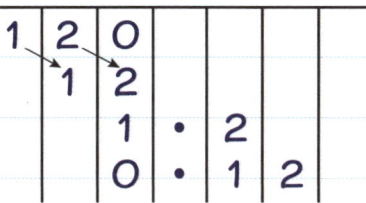

Worked example — Target grade 3

The weight of a ream of paper is 2.5 kg.
There are 500 sheets of paper in a ream.
Work out the weight, in grams, of one sheet of paper.

$2.5 \times 1000 = 2500$
$2.5\,\text{kg} = 2500\,\text{g}$
$2500 \div 500 = 5$
Each sheet of paper weighs 5 grams.

The whole ream of paper weighs 2500 g and there are 500 sheets of paper, so divide 2500 by 500 to find the weight.

Examiners' report

Students frequently lose marks by converting between different metric units incorrectly.

Start by converting 2.5 kg into grams.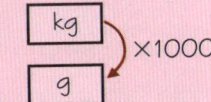

Check it!
A gram is a smaller unit than a kilogram so the number will be larger. ✓

Real students have struggled with questions like this in recent exams – **be prepared!**

Now try this

 **Target grade 2**

1 Change
 (a) 3.2 m into cm **(1 mark)**
 (b) 0.25 litres into ml **(1 mark)**
 (c) 960 mm into cm **(1 mark)**
 (d) 1700 g into kg **(1 mark)**

 **Target grade 3**

2 A pile of fifty 20p coins is 8.5 cm high.
Work out the thickness of one 20p coin. Give your answer in mm. **(2 marks)**

Worked solution video

61

RATIO & PROPORTION — Tricky Topic

Had a look ☐   Nearly there ☐   Nailed it! ☐

Reverse percentages

You can use a **multiplier** to calculate a percentage increase or decrease. If you are given the final amount and are asked to find the **original amount** you can divide by the multiplier. Here are two examples:

1 A sweater is reduced in price by 20% in a sale.

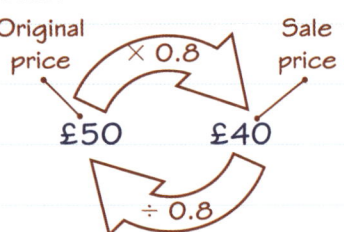

2 The average temperature increases by 5%.

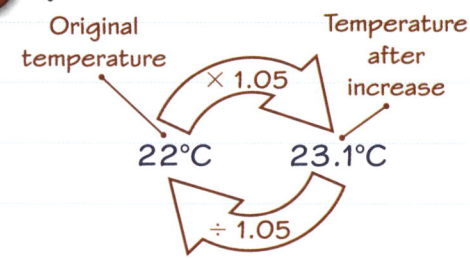

Worked example    Target grade 4

In a sale, normal prices are reduced by 15%
The sale price of a pair of trainers is £75.65
Work out the normal price of the trainers.
(3 marks)

100% − 15% = 85%

$\frac{85}{100} = 0.85$

75.65 ÷ 0.85 = 89

The original price was £89

This question tells you the price **after** the percentage decrease, so you need to use **reverse percentages**. To find the multiplier for a 15% decrease:

1. Subtract 15% from 100%
2. Divide by 100 to convert to a multiplier.

You need to **divide** by the multiplier to find the original price.

Check it!
Reduce £89 by 15%:
£89 × 0.15 = £13.35
£89 − £13.35 = £75.65 ✓

 Plan your answer before you start. You will need to do **two separate calculations** to find both original heights. You then need to **compare** the original heights and write a conclusion.

To find a multiplier for a percentage increase:
1. Add the percentage to 100%
2. Divide by 100 to convert to a multiplier.

You need to **divide** by the multiplier to find the heights in 2013.

Worked example    Target grade 4

Amy and Paul measure their height each year. This table shows their heights in 2014.

	Height in 2014 (cm)	Percentage increase since 2013
Paul	154	10%
Amy	150.8	4%

Who was taller in 2013?
Give reasons for your answer.   **(4 marks)**

Paul's original height = 154 ÷ 1.1 = 140 cm
Amy's original height = 150.8 ÷ 1.04 = 145 cm
In 2013 Amy was taller than Paul by 5 cm.

Now try this

1 Hannah buys some shoes in a sale marked '40% off'.
She pays £27 for the shoes.
What price were the shoes originally?
(3 marks)

2 Jared bought a house in 2010.
By 2012 his house had increased in value by 8%.
The new value of Jared's house is £237 600.
How much did Jared pay for his house?   **(3 marks)**

 Worked solution video    Target grade 4

Had a look ☐   Nearly there ☐   Nailed it! ☐

RATIO & PROPORTION

Tricky Topic

Growth and decay

You can use **repeated percentage change** to model problems involving growth and decay.

Compound interest

If you leave your money in a bank account it will earn compound interest.

Saanvi invests £40 000 at a compound interest rate of 3% per annum.

'Per annum' means 'per year'.

This table shows the amount of interest she earns each year.

Year	Balance (£)	Interest earned (£)
1	40 000	1200
2	41 200	1236
3	42 436	1273.08
	43 709.08	3709.08

For year 2 you have to calculate 3% of £41 200

After 3 years the total interest earned will be £3709.08 and the balance will be £43 709.08

Using index notation

This table uses a multiplier to work out the balance of Saanvi's account at the end of each year.

Year	Balance (£)
1	40 000 × 1.03 = 41 200
2	41 200 × 1.03 = 42 436
3	42 436 × 1.03 = 43 709.08

You can use indices to work out the final balance after 3 years more easily.

Balance after 3 years
= £40 000 × 1.03 × 1.03 × 1.03
= £40 000 × 1.03^3
= £43 709.08

For a reminder about working with indices look at pages 8 and 9.

Worked example   Target grade 5

At the start of an experiment a petri dish contains 5000 cells. The number of cells in the petri dish increases by 20% each day.

Calculate the number of cells in the petri dish at the end of 4 days.   **(2 marks)**

$5000 \times 1.2^4 = 10\,368$

You can use this rule to calculate a repeated percentage change.

Final amount = (starting amount) × (multiplier)^n

n is the number of times the change is made. In this example, n is the number of days.

Repeated decrease

If a question involves **depreciation** or **decay** you might need to work out a repeated percentage decrease.

This car **depreciates** in value by 8% each year.

The multiplier for an 8% decrease is 0.92

After 3 years the car is worth:

£15 000 × 0.92 × 0.92 × 0.92
= £15 000 × 0.92^3 = £11 680.32

Now try this

Worked solution video

Target grade 5

Amir invests £5000 in a savings account. He is paid 3% per annum compound interest.

(a) How much will Amir have in his savings account after 2 years?   **(2 marks)**

(b) Amir needs £5600 to buy a car. Calculate the number of years Amir would need to leave his money in the account to save up this amount.   **(2 marks)**

63

RATIO & PROPORTION | Had a look ☐ | Nearly there ☐ | Nailed it! ☐

Speed

This is the **formula triangle** for **speed**.

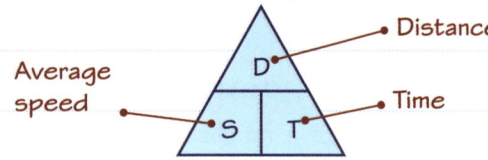
- Distance
- Average speed
- Time

Average speed = $\dfrac{\text{Total distance travelled}}{\text{Total time taken}}$

Time = $\dfrac{\text{Distance}}{\text{Average speed}}$ **LEARN IT!**

Distance = Average speed × Time

Using a formula triangle
Cover up the quantity you want to find with your finger.

The position of the other two quantities tells you the formula.

$T = \dfrac{D}{S}$ $S = \dfrac{D}{T}$ $D = S \times T$

Units
The most common units of speed are
- metres per second: m/s
- kilometres per hour: km/h
- miles per hour: mph.

The units in your answer will depend on the units you use in the formula.

When distance is measured in **km** and time is measured in **hours**, speed will be measured in **km/h**.

Minutes and hours
For questions on speed, you need to be able to convert between minutes and hours.

Hours	Minutes
$\frac{1}{2}$	30
$\frac{3}{4}$	45
$2\frac{1}{4}$	135

You can also write this as '2 hours and 15 minutes'.
60 + 60 + 15 = 135

Worked example *Target grade 3*

A plane travels at a constant speed of 840 km/h for 45 minutes.
How far has it travelled? **(3 marks)**

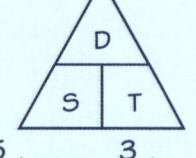

45 minutes = $\dfrac{45}{60}$ hour = $\dfrac{3}{4}$ hour

$D = S \times T$
$= 840 \times \dfrac{3}{4} = \dfrac{840 \times 3}{4} = \dfrac{2520}{4} = 630$

The plane has travelled 630 km.

Speed questions
 Draw a formula triangle.
 Make sure the units match.
 Give units with your answer.

> Make sure that the units match. Speed is given in km/h, so convert the time into hours by dividing by 60. The units of distance will be km.

Now try this *Target grade 3*

1. Bradley cycled 170 km at an average speed of 40 kilometres per hour.
   How long did it take him?
   Give your answer in hours and minutes. **(3 marks)**

2. Simon drives from Newcastle to Oxford.
   His average speed is 56 mph.
   The journey takes 4 hours 45 minutes.
   How far did he drive? **(3 marks)**

Had a look ☐   Nearly there ☐   Nailed it! ☐   **RATIO & PROPORTION**

Density

The density of a material is its mass per unit volume.

This is the formula triangle for density.

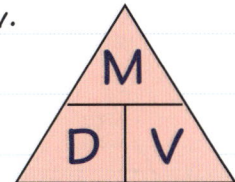

$$\text{Density} = \frac{\text{Mass}}{\text{Volume}}$$

$$\text{Volume} = \frac{\text{Mass}}{\text{Density}}$$

$$\text{Mass} = \text{Density} \times \text{Volume}$$

LEARN IT!

Worked example — Target grade 4

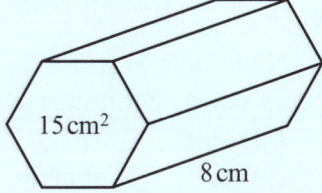

The diagram shows a solid hexagonal prism.
The area of the cross-section of the prism is 15 cm².
The length of the prism is 8 cm.
The prism is made from wood with a density of <u>0.8 grams per cm³</u>.
Work out the mass of the prism. **(4 marks)**

Volume of prism
   = Area of cross-section × Length
   = 15 × 8
   = 120 cm³

$M = D \times V$
   = 0.8 × 120
   = 96

The mass of the prism is 96 g.

Revise volumes of prisms on page 84.

Units

The most common units of density are
✓ grams per cubic centimetre: g/cm³
✓ kilograms per cubic metre: kg/m³.

Examiners' report

Make sure you write down the formula triangle for density and that you **know how to use it**. In this question you want to find the **mass**. If you cover up M the formula triangle tells you that:

Mass = Density × Volume

Real students have struggled with questions like this in recent exams – **be prepared!**

Worked example — Target grade 4

An iron bar has a volume of 1.2 m³ and a mass of 9444 kg. Calculate the density of iron.
(2 marks)

$$D = \frac{M}{V} = \frac{9444}{1.2} = 7870 \text{ kg/m}^3$$

Volume is in m³ and mass is in kg so density will be in kg/m³.

Now try this

Target grade 5

The density of copper is 8.92 g/cm³.
The density of silver is 10.49 g/cm³.
20 cm³ of copper and 5 cm³ of silver are mixed together to make a new kind of metal.
Work out the density of the new metal.
(4 marks)

Plan your answer:
1. Work out the masses of each metal.
2. Add the masses together.
3. You know the total mass and the total volume of the new metal, so calculate the density.

65

Other compound measures

Compound measures are made up of two or more other measurements. **Speed** is a compound measure because it is calculated using distance **and** time.

Pressure

Pressure is a measure of the force applied over a given area. The most common units of pressure are newtons per square centimetre (N/cm^2) and newtons per square metre (N/m^2).

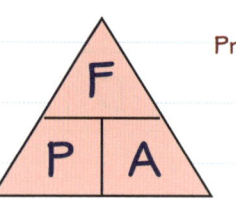

$$\text{Pressure} = \frac{\text{Force}}{\text{Area}}$$

$$\text{Area} = \frac{\text{Force}}{\text{Pressure}}$$

$$\text{Force} = \text{Pressure} \times \text{Area}$$

LEARN IT!

Worked example — Target grade 4

At a depth of 15 m, water has a pressure of 14.7 N/cm^2. Calculate the force applied to a diving mask with a surface area of 360 cm^2. **(2 marks)**

Force = Pressure × Area
= 14.7 × 360 = 5292 N

Rates

If the bottom unit in a compound measure is **time**, then it is a **rate**.

$$\text{Speed} = \frac{\text{Distance}}{\text{Time}} \qquad \text{Rate of climb} = \frac{\text{Height}}{\text{Time}}$$

$$\text{Rate of flow} = \frac{\text{Volume}}{\text{Time}} \qquad \text{Rate of pay} = \frac{\text{Salary}}{\text{Time}}$$

Problem solved! You need to use lots of different maths skills to solve this problem. You need to know how to calculate the volume of a cuboid (revise this on page 83) and how to convert from cm^3 to litres (for a reminder look at page 61). It's a good idea to **plan your strategy** before you start:

1. Work out the volume of the fishtank in litres.
2. Divide by 2 to get half the capacity.
3. Divide by the rate of flow to get the time taken in minutes.

Worked example — Target grade 5

This fishtank can be modelled as a cuboid.

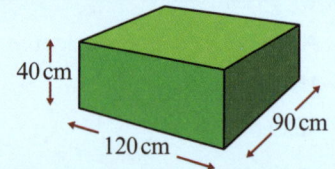

Aaron fills the fishtank with water at a rate of 12 litres per minute. How long will it take for the fishtank to be half full? **(5 marks)**

Volume = 120 × 90 × 40
= 432 000 cm^3 = 432 litres

432 ÷ 2 = 216   216 ÷ 12 = 18

It will take 18 minutes to half fill the tank.

Now try this

 **1** The average fuel consumption of a car is measured in kilometres per litre (km/litre). A car travels 249 km and uses 15 litres of petrol. What is its average fuel consumption? **(2 marks)**

Look at the units to work out what calculation to do.

 **2** A large tank holds 1680 litres of water. The tank can be filled from a hot tap or from a cold tap. The cold tap on its own takes 4 minutes to fill the tank. The hot tap on its own takes 6 minutes to fill the tank. Evie turns both taps on at the same time. How long does the tank take to fill? **(3 marks)**

Had a look ☐   Nearly there ☐   Nailed it! ☐   **RATIO & PROPORTION**

Proportion

Two quantities are in **direct proportion** when both quantities increase at the same rate.

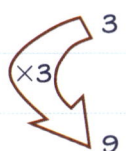
Number of tickets bought

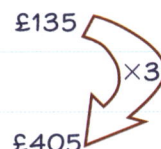
Total cost

Two quantities are in **inverse proportion** when one quantity increases at the same rate as the other quantity decreases.

Average speed   Time taken

Worked example
Target grade 2

Suresh buys 4 picture frames for a total cost of £11.40.
Work out the cost of 7 of these picture frames. **(2 marks)**

Cost of 1 frame = $\frac{£11.40}{4}$ = £2.85

Cost of 7 frames = £2.85 × 7 = £19.95

Calculate the cost of 1 picture frame first. Then multiply the cost of 1 frame by 7 to work out the cost of 7 frames. When you are working with money you should:
- do all your calculations in either £ or p
- write £ or p in your answer, but not both
- write answers in £ to 2 decimal places.

Multiply or divide?

6 people can build a wall in 4 days.
How long would it take 8 people to build the same wall?

Inverse proportion problems often involve time. The more people working on a task, the quicker it will be finished.

You can solve this problem by working out how long it would take 1 person to build the wall. Use common sense to decide whether to divide or multiply.

6 × 4 = 24 so 1 person could build the wall in 24 days.
You multiply because it would take 1 person more time to build the wall.

24 ÷ 8 = 3 so 8 people could build the wall in 3 days.
You divide because it would take 8 people less time to build the wall.

Now try this

Target grade 2

1 Eight identical bottles of water cost £4.48.
   (a) Work out the cost of 5 of these bottles of water. **(2 marks)**
   (b) Work out the cost of 12 of these bottles of water. **(1 mark)**

Worked solution video

Target grade 3

2 It takes 8 men a total of 6 days to dig a hole.
   How long would it take 3 men to dig a hole of the same size? **(2 marks)**

Work out how long it would take one man to dig the hole. It will take longer, so multiply.

RATIO & PROPORTION

Had a look ☐   Nearly there ☐   Nailed it! ☐

Tricky Topic

Proportion and graphs

You can use the symbol ∝ to show direct proportion. You can show quantities that are **directly proportional** or **inversely proportional** on a graph.

Direct proportion facts

If y is directly proportional to x:

- ✓ You can write $y \propto x$
- ✓ You can write an equation $y = kx$ where k is a number.
- ✓ The graph of x against y is a **straight line** passing **through the origin**.

Inverse proportion facts

If y is inversely proportional to x:

- ✓ You can write $y \propto \dfrac{1}{x}$ — *y is directly proportional to the **reciprocal** of x*
- ✓ You can write an equation $y = \dfrac{k}{x}$ where k is a number.
- ✓ The graph of x against y looks like a **reciprocal graph**.

Worked example (Target grade 5)

p and q are inversely proportional. Circle the equation that could describe the relationship between p and q.

$p = 2q \quad p = q + 5 \quad p = \dfrac{q}{10} \quad \boxed{p = \dfrac{2}{q}}$ **(1 mark)**

> An equation for **inverse** proportionality looks like $y = \dfrac{k}{x}$ where k is a number.

> There are other ways to answer this question. The method used here is a bit like using **equivalent ratios**:
>
>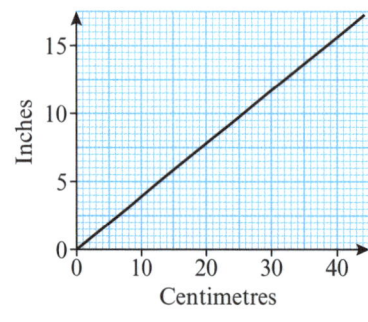
>
> **Check it!**
> Your answer should make sense. If the quantities are in **direct proportion** then when one decreases the other should decrease as well.

Worked example (Target grade 5)

Electrical appliances in your home follow this rule:

$$\text{Power (watts)} \propto \text{Current (amps)}$$

An electric drill uses 2.2 A and has a power of 528 W. Calculate the current used by a television with a power of 132 W.   **(2 marks)**

$528 \div 132 = 4$
$2.2 \div 4 = 0.55$
The television uses 0.55 A

Now try this

This graph can be used to convert between inches and centimetres.

[Graph: Inches (y-axis, 0 to 15) vs Centimetres (x-axis, 0 to 40), straight line through origin]

(a) Use the graph to convert 10 cm to inches. **(1 mark)** *(Target grade 4)*

(b) Use the graph to convert 12.5 inches to cm. **(1 mark)** *(Target grade 4)*

(c) What evidence is there from the graph to show that inches are directly proportional to centimetres? **(2 marks)** *(Target grade 5)*

Had a look ☐   Nearly there ☐   Nailed it! ☐

RATIO & PROPORTION

Problem-solving practice 1

About half of the questions in your Foundation GCSE exam will require you to **problem-solve**, **reason**, **interpret** or **communicate** mathematically. If you come across a tricky or unfamiliar question in your exam you can try some of these strategies:
- ✓ Sketch a diagram to see what is going on.
- ✓ Try the problem with smaller or easier numbers.
- ✓ Plan your strategy before you start.
- ✓ Write down any formulae you might be able to use.
- ✓ Use x or n to represent an unknown value.

AO2
AO3

Now try this

1 A bag contains green, red and blue counters.
20% of the counters are green.
There are three times as many red counters as blue counters.
There are 9 blue counters in the bag.
How many counters are there in the bag in total? **(4 marks)**

Percentages page 55 *Target grade 2*

There is more than one way to approach this question. One way is to start by working out how many red and blue counters there are in total. This represents 80% of all the counters in the bag.

TOP TIP

You can sometimes solve percentage problems by working out what 1% represents and then multiplying by 100.

---

2 Suresh wants to buy a new pair of trainers.
There are three shops that sell the trainers he wants.

Sportcentre Trainers	Footwear First Trainers	Action Sport Trainers
£10 plus 10 payments of £3.50	$\frac{1}{4}$ off usual price of £80	£40 plus VAT at 20%

Which shop is selling the trainers the cheapest?
You must show working to justify your answer. **(5 marks)**

Fractions, decimals and percentages page 56
Percentage change 1 page 57 *Target grade 3*

There are lots of steps in this question so make sure you keep track of your working. You need to calculate the price at each shop, then write down which shop is cheapest.

TOP TIP

Divide the answer space into three columns. Then the examiner can see which shop each bit of working is for.

Problem-solving practice 2

3 At a sixth form college, the ratio of male to female students is 7 : 6
There are 144 female students at the college. How many students are there in total? **(2 marks)**

Ratio 2 page 60 — Target grade 4
Use the information in the question to work out what each part of the ratio represents. There are 7 + 6 = 13 parts in the ratio, so multiply your answer by 13 to work out the total number of students.

TOP TIP

Check your answer by dividing it in the ratio 7 : 6

---

4 Joshua invested £2000 in a savings account for two years. The account pays 2% per annum compound interest.
At the end of the first year he put an extra £1500 into the account.
How much money will Joshua have at the end of the two years? **(4 marks)**

Growth and decay page 63 — Target grade 5
Work out the interest Joshua earns on £2000, then add the extra £1500. This is the new starting amount for the second year. So in the second year, Joshua earns interest on this new amount.

TOP TIP

At each step of your working, make a note of what you have worked out. Your notes here could be:
- Interest earned in year 1
- Total starting amount for year 2
- Interest earned in year 2
- Total amount at end of year 2

---

5 This item appeared in a newspaper.

> **Cow produces 3% more milk**
> A farmer found that when his cow listened to classical music the milk it produced increased by 3%.
> This increase of 3% represented 0.72 litres of milk.

Calculate the amount of milk produced by the cow when it listened to classical music. **(4 marks)**

Proportion page 67 — Target grade 5
When the cow listened to classical music, it produced 103% of the milk it produced originally. You know that 3% represents 0.72 litres. Use this information to work out what 103% represents.

TOP TIP

You can sometimes solve percentage problems by working out what 1% represents.

Had a look ☐   Nearly there ☐   Nailed it! ☐   **GEOMETRY & MEASURES**

Symmetry

Lines of symmetry

A line of symmetry is a mirror line. One half of the shape is a mirror image of the other.

No lines of symmetry

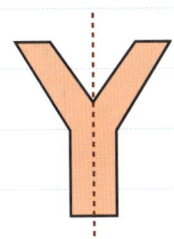
1 line of symmetry

Using tracing paper

You are allowed to ask for tracing paper in your exam. You can use it to check for lines of symmetry.

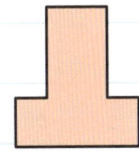 

If you fold a tracing of a shape in half along a line of symmetry, the two halves will match up exactly.

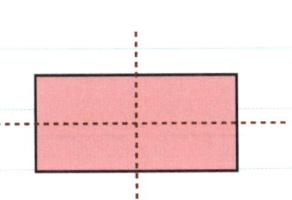
2 lines of symmetry

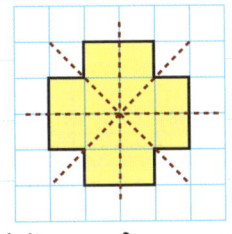
4 lines of symmetry

Rotational symmetry

If a shape fits over itself when it is rotated then it has rotational symmetry. The order of rotational symmetry tells you how many times it fits over itself in one full turn.

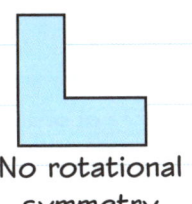
No rotational symmetry

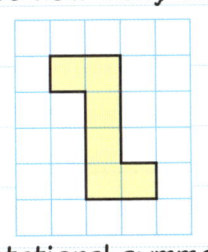
Rotational symmetry of order 2

Using tracing paper

You can use tracing paper to check for rotational symmetry.

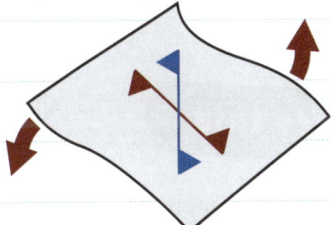

Trace the shape. Rotate the tracing paper and see how many times the shape fits over itself.

This shape fits over itself twice: once at 180° and once at 360°. It has rotational symmetry of order 2.

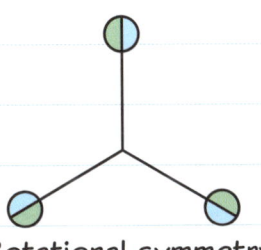
Rotational symmetry of order 3

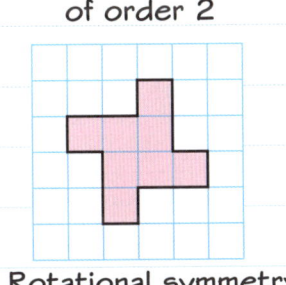
Rotational symmetry of order 4

Now try this

 **Target grade 1**
(a) Fill in 5 squares of the grid below to make a shape with exactly 4 lines of symmetry.

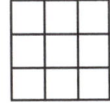

(1 mark)

 **Target grade 2**
(b) Fill in 5 squares of the grid below to make a shape with rotational symmetry of order 2.

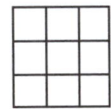

(1 mark)

71

GEOMETRY & MEASURES   Had a look ☐   Nearly there ☐   Nailed it! ☐

Quadrilaterals

A **quadrilateral** is any four-sided shape. The lines joining the opposite corners of a quadrilateral are called **diagonals**.

Square 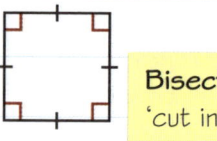 *Bisect means 'cut in half exactly'.* All sides equal. ✓ All angles are 90°. ✓ Diagonals are equal and bisect each other at 90°. ✓ 4 lines of symmetry. ✓ Rotational symmetry of order 4. ✓	**Rectangle** 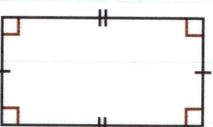 Opposite sides equal. ✓ All angles are 90°. ✓ Diagonals are equal and bisect each other. ✓ 2 lines of symmetry. ✓ Rotational symmetry of order 2. ✓	**Parallelogram** 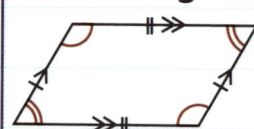 Opposite sides parallel and equal. ✓ Opposite angles equal. ✓ Diagonals bisect each other. ✓ No lines of symmetry. ✓ Rotational symmetry of order 2. ✓
Trapezium 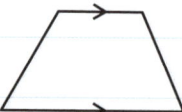 One pair of opposite sides parallel. ✓ *Matching arrows mean lines are **parallel**. Matching dashes mean lines are **the same length**.*	**Kite** 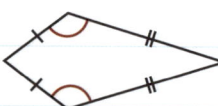 Two pairs of adjacent sides equal. ✓ One pair of opposite angles equal. ✓ Diagonals cross at 90°. ✓ 1 line of symmetry. ✓	**Rhombus** 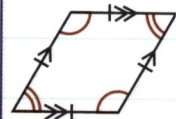 All sides equal. ✓ Opposite sides parallel. ✓ Opposite angles equal. ✓ 2 lines of symmetry. ✓ Diagonals bisect each other at 90°. ✓ Rotational symmetry of order 2. ✓

Now try this

Target grade 1

Join the name of each 2-D shape to its correct description with a straight line.
The first one has been done for you.   **(5 marks)**

Shape	Description
Square	Diagonals cross at 90° to each other. Rotational symmetry of order 2.
Rectangle	Diagonals bisect each other but are not at 90°. No lines of symmetry.
Rhombus	One pair of parallel sides. No rotational symmetry.
Kite	All angles 90°. Diagonals cross at 90° to each other.
Trapezium	Diagonals cross at 90° to each other. One line of symmetry.
Parallelogram	Diagonals bisect each other but are not at 90°. Two lines of symmetry.

Had a look ☐ Nearly there ☐ Nailed it! ☐   **GEOMETRY & MEASURES**

Angles 1

Types of angle

You need to know the names of the different types of angles.

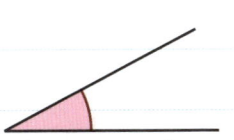

Acute angle: Less than 90°

Right angle: 90°

Obtuse angle: Between 90° and 180°

Reflex angle: More than 180°

You can use these angle types to help you estimate the size of angles.

To revise measuring and drawing angles, have a look at page 95.

Naming angles

You can use letters to name angles.

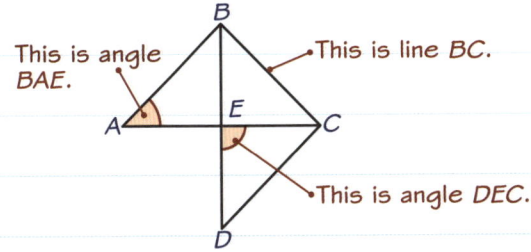

This is angle BAE.
This is line BC.
This is angle DEC.

Angles are named using the three letters of the lines that make the angle. The angle is always at the middle letter.

Special triangles

Here are three special types of triangle.

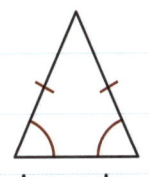 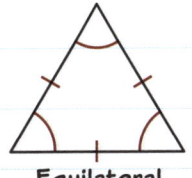 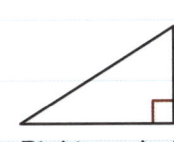

Isosceles — Two equal sides, Two equal angles

Equilateral — Three equal sides, All angles 60°

Right-angled — One angle 90°

In a **scalene** triangle none of the sides or angles are equal.

Angle facts

You can use these angle facts to work out missing angles.

1 Angles on a straight line add up to 180°.

$a + b = 180°$

2 Angles around a point add up to 360°.

$c + d + e = 360°$

3 Vertically opposite angles are equal.

$f = g$

$h = i$

Worked example *Target grade 1*

(a) Mark with a letter R a reflex angle. **(1 mark)**

(b) Mark with a letter O an obtuse angle. **(1 mark)**

(c) Work out the size of angle p. Give a reason for your answer. **(2 marks)**

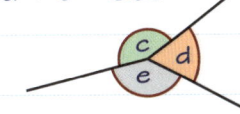

$p + 52° = 180°$
$180 - 52 = 128$
$p = 128°$

Angles on a straight line add up to 180°.

Everything in blue is part of the answer.

Now try this *Target grade 1*

Work out the sizes of the angles marked with letters in this diagram. Give reasons for your answers. **(3 marks)**

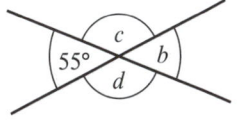

Worked solution video

You could start by using angle fact 1 or angle fact 3 from the blue box above.

73

Angles 2

Triangles and quadrilaterals

These are useful facts for triangles and quadrilaterals.

1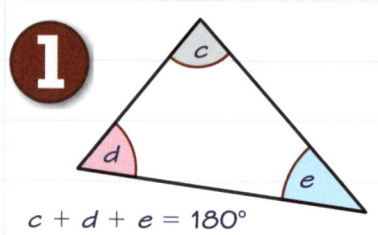
$c + d + e = 180°$
Angles in a triangle add up to 180°.

2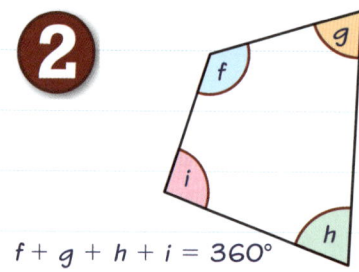
$f + g + h + i = 360°$
Angles in a quadrilateral add up to 360°.

3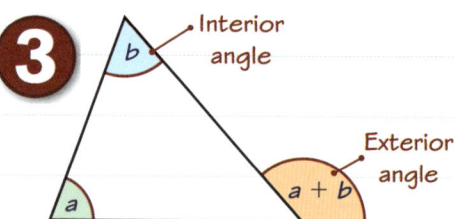
The exterior angle of a triangle is equal to the sum of the interior angles at the other two vertices.

4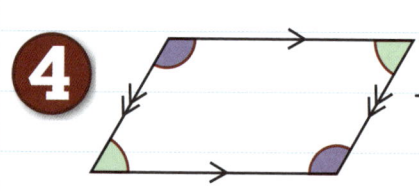
The opposite angles of a parallelogram are equal.

Parallel and perpendicular lines

Perpendicular lines meet at 90°.
Lines that remain the same distance apart are **parallel**.
Parallel lines are marked with arrows.
You need to remember these angle facts about parallel lines and their correct names:

1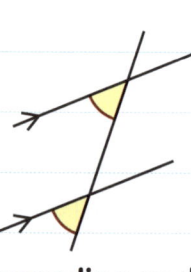
Corresponding angles are equal.

2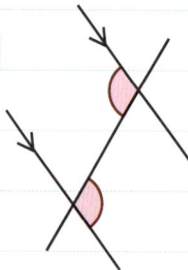
Alternate angles are equal.

3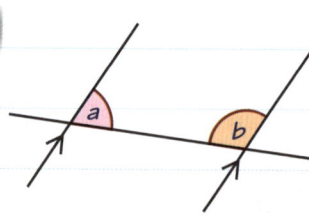
$a + b = 180°$
Co-interior or **allied** angles add up to 180°.

Worked example — Target grade 3

Work out the size of the angle marked x.
Show clearly, giving reasons, how you work out your answer. **(3 marks)**

Angle $CGH = 118°$
Corresponding angles are equal.
$180 - 118 = 62$
Angle $x = 62°$
Angles on a straight line add up to 180°.

Now try this — Target grade 3

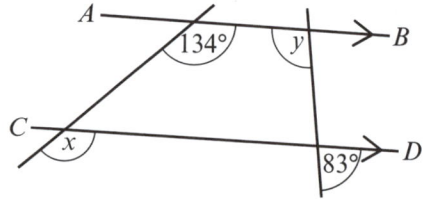

(a) Write down the size of angle x.
Give a reason for your answer. **(2 marks)**
(b) Work out the size of angle y. **(2 marks)**

Write any angles you have worked out on the diagram.

Had a look ☐   Nearly there ☐   Nailed it! ☐

GEOMETRY & MEASURES

Solving angle problems

When you are solving angle problems you need to give reasons for each step of your working. Look at pages 73 and 74 to see the reasons you can use to solve angle problems.

Worked example *Target grade 3*

Find the value of x.

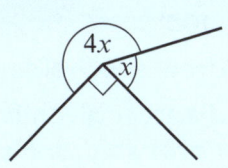

(3 marks)

$360° − 90° = 270°$

$4x + x = 5x$

$5x = 270°$

$x = 54°$

Examiners' report

✗ **Trial and improvement** methods are really hit-and-miss. And if you don't find the correct answer, you're likely to get no marks.

✗ Don't attempt to **measure** the angle. Unless you're told otherwise, diagrams are not drawn to scale in your exam.

✓ If you attempt to use **algebra** for this question you are much more likely to pick up a method mark.

Real students have struggled with questions like this in recent exams – **be prepared!**

There is more about simplifying algebraic expressions like this on page 22.

Worked example *Target grade 3*

The arrows tell you which lines are parallel.
The dashes tell you which lines are equal.

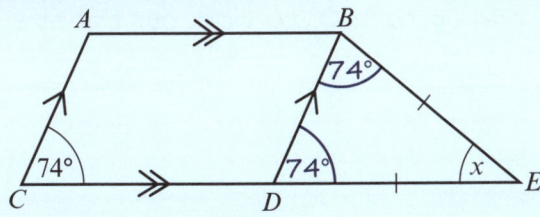

Work out the value of x.
Show clearly, giving reasons, how you work out your answer. **(4 marks)**

Everything in blue is part of the answer.

Angle $BDE = 74°$
Corresponding angles are equal.
Angle $DBE = 74°$
Base angles of an isosceles triangle are equal.
$180 − 74 − 74 = 32$
$x = 32°$
Angles in a triangle add up to $180°$.

Now try this

Remember to give reasons for each step of your working.

Target grade 3

1. Work out the size of
   (a) angle GCD **(1 mark)**
   (b) angle FCG. **(2 marks)**

   Diagram: A with $71°$ at C, B---D horizontal line, $95°$ at E, F, G, H

Worked solution video

2. In the diagram AB is parallel to CD.

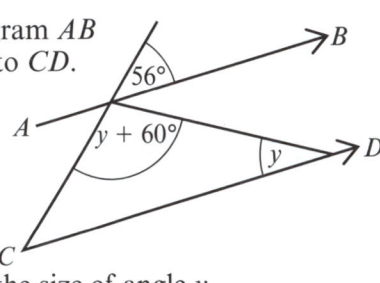

Work out the size of angle y.
You must show your working. **(4 marks)**

75

GEOMETRY & MEASURES

Had a look ☐   Nearly there ☐   Nailed it! ☐

Tricky Topic

Angles in polygons

Polygon questions are all about interior and exterior angles.

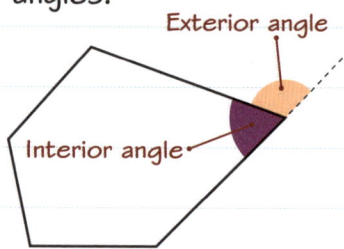

This diagram shows part of a **regular** polygon with 30 sides.

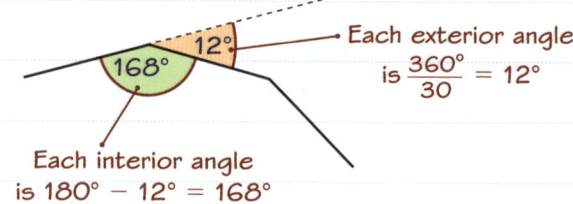

Each exterior angle is $\frac{360°}{30} = 12°$

Each interior angle is $180° - 12° = 168°$

Use these formulae for a polygon with n sides:
Sum of interior angles = $180° \times (n - 2)$
Sum of exterior angles = $360°$

Don't try to draw a 30-sided polygon! If there's no diagram given in a polygon question, you probably don't need to draw one.

Regular polygons

In a regular polygon all the sides are equal and all the angles are equal.
If a regular polygon has n sides then each exterior angle is $\frac{360°}{n}$

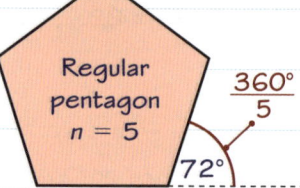 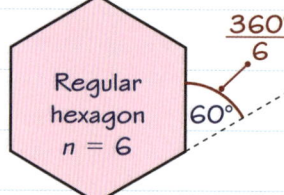 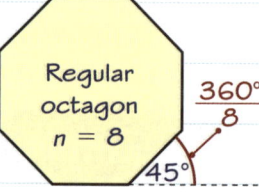

You can use the fact that angles on a straight line add up to 180° to work out the size of one of the interior angles.

Worked example *Target grade 4*

The diagram shows part of a regular polygon.
Work out the number of sides in the polygon.

(3 marks)

Exterior angle = $180° - 156° = 24°$
$\frac{360°}{n}$ so $n = \frac{360°}{24°} = 15$
The polygon has 15 sides.

Problem solved! It's usually easier to work with **exterior angles** than interior angles. Use the fact that angles on a straight line add up to 180° to calculate the size of one exterior angle.

In a regular n-sided polygon, each exterior angle is $\frac{360°}{n}$. You can use this fact to write an equation and solve it to find n.

Now try this *Target grade 4*

JK, KL and LM are three sides of a 20-sided regular polygon.
JK and ML are extended to meet at T.
Work out the size of angle KTL, marked x on the diagram.   (4 marks)

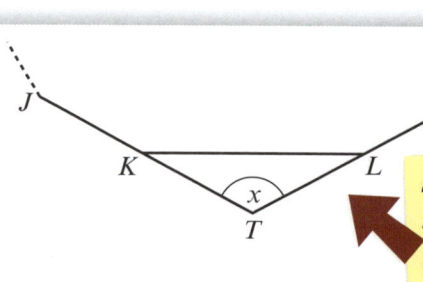

Angles TKL and TLK are both external angles of the 20-sided polygon.

Had a look ☐   Nearly there ☐   Nailed it! ☐   **GEOMETRY & MEASURES**

Time and timetables

Time

1 You can write time using the 12-hour clock or the 24-hour clock.

12-hour clock	24-hour clock
8.15 am	08:15
4.50 pm	16:50
12.00 midday	12:00
12.00 midnight	00:00

2 Remember there are 60 minutes in an hour.

Hours	Minutes
$\frac{1}{2}$	30
$\frac{3}{4}$	45
$2\frac{1}{4}$	135

You can also write this as '2 hours and 15 minutes'.
$60 + 60 + 15 = 135$

Timetables

This table shows part of a bus timetable.
Timetables usually give times using the 24-hour clock.

Crook	08:15	09:15	10:45	11:15
Prudhoe	08:28	09:28	10:58	11:28
Hexham	08:45	09:45	11:15	11:45
Alton	09:00	10:00	11:30	12:00

This bus leaves Crook at 10:45 and arrives in Hexham at 11:15

This bus leaves Prudhoe at 11:28 and arrives in Alton at 12:00
11:28 to 11:30 = 2 minutes
11:30 to 12:00 = 30 minutes
The journey time is 32 minutes.

Worked example   Target grade 1

Olivia is visiting her sister in Exeter. She leaves home at 11.30 am. She writes down how long each section of her journey takes.

Drive to train station   15 minutes
Wait for train   25 minutes
Train journey $1\frac{1}{2}$ hours

At what time does Olivia arrive in Exeter?

11.30 am → 11.45 am → 12.10 pm → 1.40 pm

Olivia arrives at 1.40 pm.

Add on each time in steps. Write down each new time.

11.30 am +15 min→ 11.45 am +25 min→ 12.10 pm
(+40 min)

12.10 pm +1 hour→ 1.10 pm +30 min→ 1.40 pm
(+$1\frac{1}{2}$ hours)

Now try this   Target grade 1

The clocks show the time in London and San Francisco.
It is afternoon in London.
(a) Write down the time in London using the 24-hour clock. **(1 mark)**
It is morning in San Francisco when it is afternoon in London.
The time in London is ahead of the time in San Francisco.
(b) How many hours is London ahead of San Francisco? **(1 mark)**

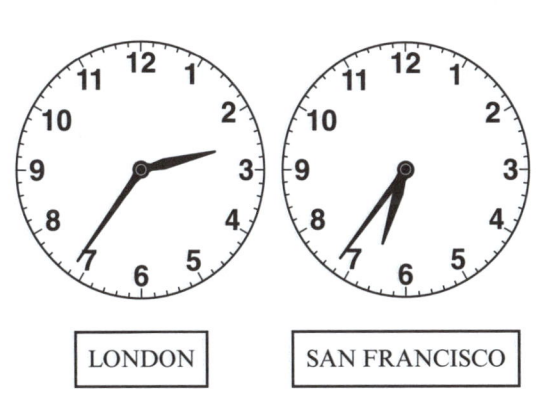

LONDON   SAN FRANCISCO

GEOMETRY & MEASURES Had a look ☐   Nearly there ☐   Nailed it! ☐

Reading scales

Here are three things you need to watch out for when reading scales.

1 You need to be able to read scales and number lines. Begin by working out what each division on a scale represents.

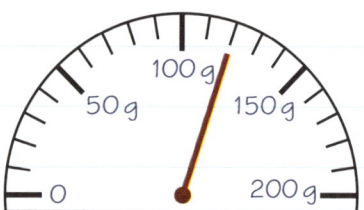

There are 5 divisions between 100 g and 150 g.
Each division represents 10 g.
The scale reads 120 g.

2 Not all divisions represent 1 unit or 10 units.

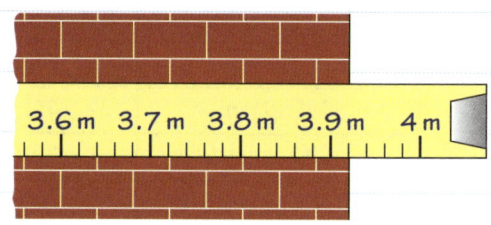

There are 5 divisions between 3.9 m and 4 m.
$0.1 \div 5 = 0.02$
Each division represents 0.02 m.
This wall is 3.92 m long.

Worked example  Target grade 1

The diagram below shows 4 identical cubes and 4 identical tetrahedrons.

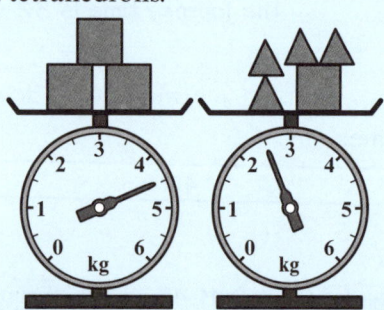

Work out the weight in kg of one tetrahedron.
(3 marks)

$4.5 \div 3 = 1.5$ so 1 cube = 1.5 kg
$2.5 - 1.5 = 1$
$1 \div 4 = 0.25$ so 1 tetrahedron = 0.25 kg

3 Sometimes you have to **estimate** the reading on a scale.

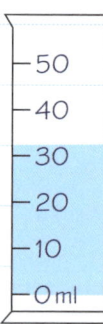

The water doesn't come up to an exact mark but you can make an estimate.
The water is closer to 30 ml than 40 ml.
32 ml would be a good estimate.

 Plan your strategy before you start.
1. Divide the total on the first scale by 3 to work out the weight of one cube.
2. Subtract that from the reading on the second scale.
3. Divide the remainder by 4 to work out the weight of one tetrahedron.

Make sure you write down your working so you can show the strategy you used.

Now try this Target grade 1

This scale has 3 identical spheres on it.

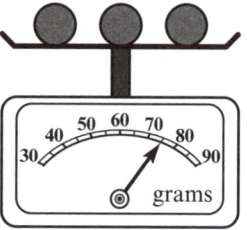

Work out the weight in grams of one sphere.   **(2 marks)**

Had a look ☐ Nearly there ☐ Nailed it! ☐ **GEOMETRY & MEASURES**

Perimeter and area

Perimeter

Perimeter is the distance around the edge of a shape. You can work out the perimeter of a shape by adding up the lengths of the sides.

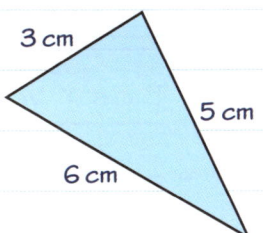

Perimeter = 3 cm + 5 cm + 6 cm
= 14 cm

You might need to measure a shape to find the perimeter.

See page 96 for help on measuring lines.

Worked example

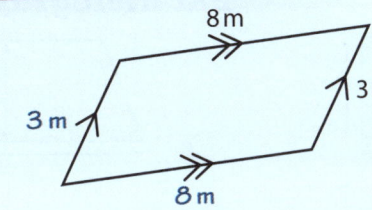

Work out the perimeter of this parallelogram. **(2 marks)**

3 + 8 + 3 + 8 = 22

Perimeter = 22 m

Everything in blue is part of the answer.

Work out the missing lengths first. The opposite sides of a parallelogram are equal so you can fill in these lengths on the diagram.

Area

You can work out the **area** of a shape drawn on squared paper by counting the squares.

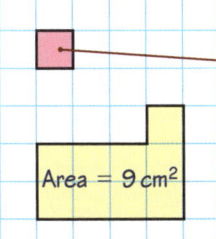

This area is 1 cm²
You say 'one centimetre squared' or 'one square centimetre'.

Worked example

This shape is drawn on cm squared paper.
(a) Work out the perimeter of the shape. **(1 mark)**

18 cm

(b) Work out the area of the shape. **(1 mark)**

12 cm²

Estimating

You might need to estimate the area of a shape drawn on cm squared paper. Count 1 cm² for every whole square and ½ cm² for every part square.

Here there are 10 whole squares and 6 part squares. A good estimate is 13 cm².

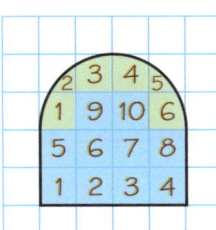

Now try this

Estimate the area of this oval shape. Each square represents 1 cm². **(2 marks)**

Count 1 cm² for each whole square and ½ cm² for each part square.

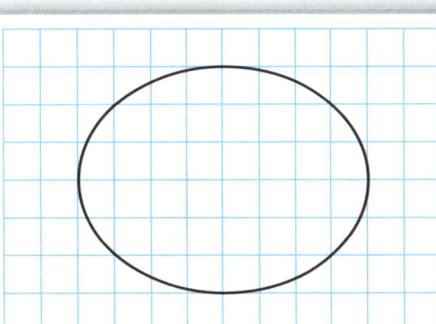

GEOMETRY & MEASURES

Had a look ☐   Nearly there ☐   Nailed it! ☐

Area formulae

You will not be given any of the formulae on this page in your exam. Make sure you **learn** them and know how to use them.

Rectangle

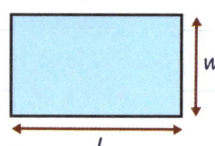

Area = Length × Width
$A = lw$

Parallelogram

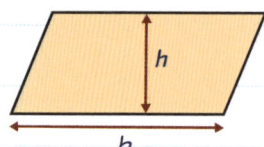

Area = Base × Vertical height
$A = bh$

Triangle

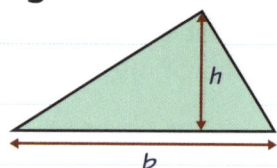

Area = $\dfrac{\text{Base} \times \text{Vertical height}}{2}$

$A = \frac{1}{2}bh$

h is the vertical height.

Trapezium

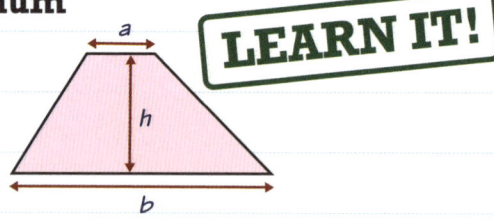

Area = $\frac{1}{2}\begin{pmatrix}\text{Sum of the two}\\ \text{parallel sides}\end{pmatrix} \times$ Vertical height

$A = \frac{1}{2}(a + b)h$

LEARN IT!

Area checklist
- ✓ Make sure lengths are all in the same units.
- ✓ Remember to give units with your answer.

Lengths in cm then area in cm^2.
Lengths in m then area in m^2.

Worked example — Target grade 3

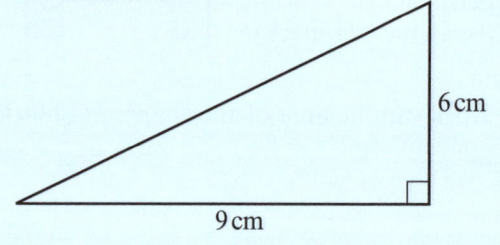

Work out the area of this triangle. **(2 marks)**

$A = \frac{1}{2}bh$
$= \frac{1}{2} \times 9 \times 6$
$= 27\ cm^2$

Multiplying by $\frac{1}{2}$ is the same as dividing by 2.

Worked example — Target grade 3

The diagram shows a rug in the shape of a trapezium. Work out the area of the rug. **(3 marks)**

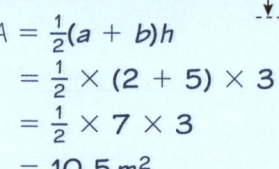

$A = \frac{1}{2}(a + b)h$
$= \frac{1}{2} \times (2 + 5) \times 3$
$= \frac{1}{2} \times 7 \times 3$
$= 10.5\ m^2$

Learn the formula for the area of a trapezium. You should always **write down** the formula before you substitute any values.

Now try this — Target grade 3

Here is a diagram of a parallelogram and a rectangle.
The parallelogram has the same area as the rectangle.
Work out the length x. **(4 marks)**

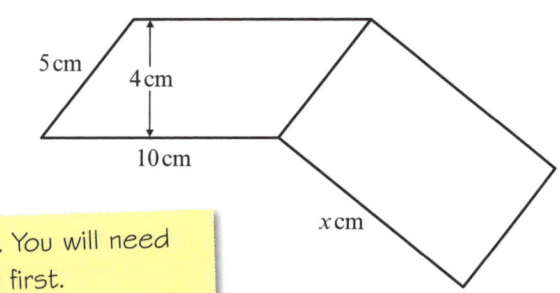

Plan your answer before you start writing. You will need to work out the area of the parallelogram first.
You can write any lengths you find on your diagram.

Had a look ☐ Nearly there ☐ Nailed it! ☐

GEOMETRY & MEASURES

Solving area problems

You can calculate areas and perimeters of harder shapes by splitting them into parts. You might need to draw some extra lines on your diagram and add or subtract areas.

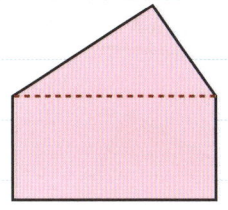
Area = Rectangle + Triangle

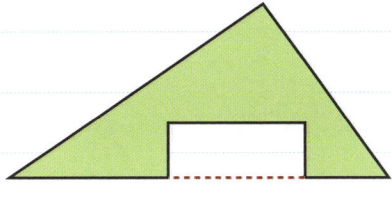
Area = Triangle − Rectangle

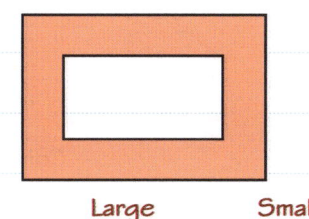
Area = Large rectangle − Small rectangle

Worked example — Target grade 3

The diagram shows a garden bed.

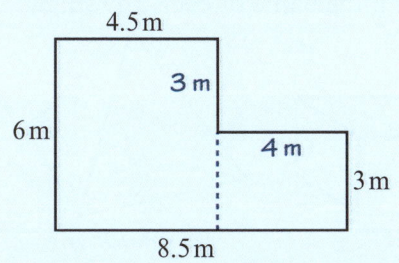

Adrian wants to cover the bed with grass seed. A packet of grass seed will cover $10\,m^2$.

(a) How many packets of grass seed does Adrian need to buy? **(3 marks)**

Area = $6 \times 4.5 + 4 \times 3$
= $27 + 12$
= $39\,m^2$

Adrian needs to buy 4 packets of grass seed.

Adrian also wants to build a fence around the edge of the garden bed.

(b) Calculate the total length of Adrian's fence. **(1 mark)**

$6 + 4.5 + 3 + 4 + 3 + 8.5 = 29\,m$

Examiners' report

Read the question carefully to decide whether you need to find an **area** or a **perimeter**. Start by finding missing lengths and **write them on the diagram**.

Real students have struggled with questions like this in recent exams – **be prepared!**

Everything in blue is part of the answer.

Draw a dotted line to divide the garden bed into two rectangles.

You have to use the information in the question to work out the missing lengths.

$8.5\,m − 4.5\,m = 4\,m$

$6\,m − 3\,m = 3\,m$

Make sure you answer the question that has been asked.

For part (a) you need to say how many packets of grass seed Adrian needs to buy.

Don't measure the distances. Unless you are told otherwise, all the diagrams in your exam are not drawn accurately.

Now try this — Target grade 3

Here is a shape made up of rectangles.

(a) Work out the distance marked x on the diagram. **(1 mark)**
(b) Work out the distance marked y on the diagram. **(1 mark)**
(c) Work out the area of this shape. **(3 marks)**
(d) Work out the perimeter of this shape. **(2 marks)**

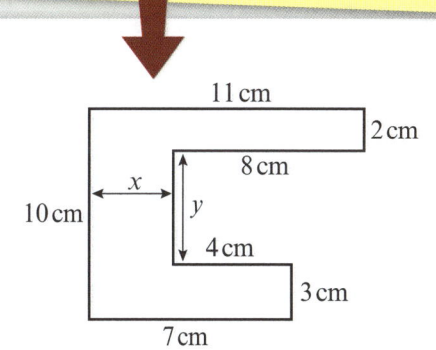

GEOMETRY & MEASURES — Had a look ☐ Nearly there ☐ Nailed it! ☐

3-D shapes

You need to learn the names of these 3-D shapes.

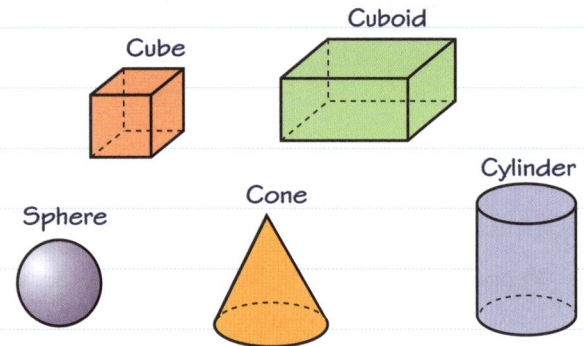
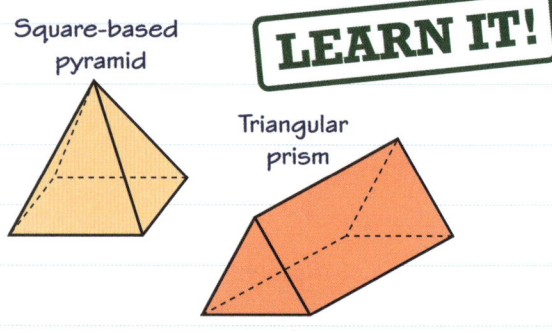

LEARN IT!

Faces, edges and vertices

This square-based pyramid has 5 faces, 8 edges and 5 vertices.

The plural of vertex is vertices.

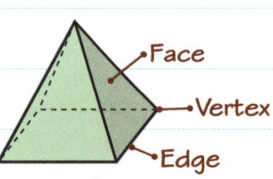

You need to **learn** the mathematical names of 3-D solids. You can sketch the hidden edges to help you work out the answer to part (b).

Worked example *Target grade 1*

(a) Write down the name of each 3-D shape.
   (i) 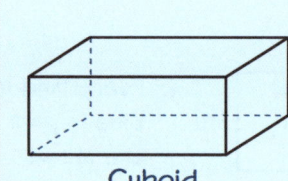 (ii)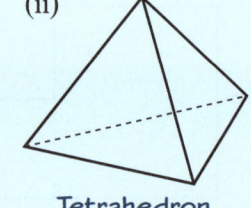
   Cuboid   Tetrahedron

(b) Write down the number of edges on shape (i).

12

Everything in blue is part of the answer.

Surface area of a cuboid

You can find the surface area of a 3-D shape by adding together the areas of its faces. A cuboid has **six** faces. You can use the fact that **opposite faces** are **equal** to simplify your working.

Surface area = $2A + 2B + 2C$
= $2 \times (6 \times 3) + 2 \times (6 \times 4) + 2 \times (4 \times 3)$
= $2 \times 18 + 2 \times 24 + 2 \times 12 = 108\,cm^2$

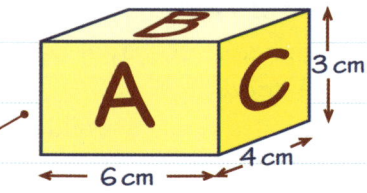

Now try this *Target grade 3*

Worked solution video

The diagram shows a solid block of wood in the shape of a cuboid.
(a) Work out the total surface area of the cuboid.   **(2 marks)**
Karl has to paint all 6 faces of 50 cuboids.
A can of spray paint covers an area of 3000 cm².
(b) How many cans of paint must Karl buy to paint all the faces? Show all your working.   **(3 marks)**

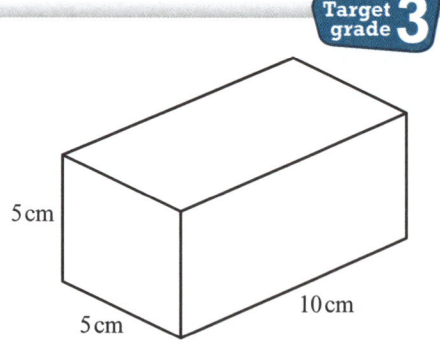

Don't just write down a number or estimate. You must show all your working.

Had a look ☐   Nearly there ☐   Nailed it! ☐   **GEOMETRY & MEASURES**

Volumes of cuboids

The **volume** of a 3-D shape is the amount of space it takes up.

The most common units of volume are cm³ or m³.

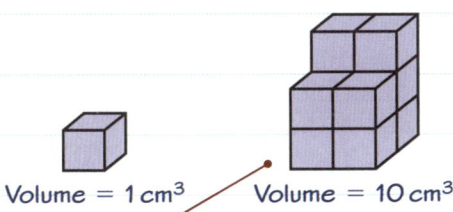

Volume = 1 cm³   Volume = 10 cm³

This shape is made from ten 1 cm³ cubes.

Volume of a cuboid

You need to remember this formula for the volume of a **cuboid**.

Volume = Length × Width × Height

Worked example — Target grade 3

The diagram shows a wooden planting box.

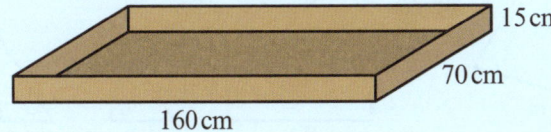

15 cm, 70 cm, 160 cm

A 50 litre bag of compost costs £3.99.
1 litre = 1000 cm³
How much will it cost to fill this planting box? **(4 marks)**

Volume = 160 × 70 × 15
      = 168 000 cm³
      = 168 litres

4 bags of compost are needed.

4 × £3.99 = £15.96

Problem solved! There are a few steps to this problem so it's a good idea to plan your answer:

1. Work out the volume of the planting box.
2. Convert from cm³ to litres by dividing by 1000.
3. Work out how many bags of compost are needed.
4. Work out how much the compost will cost.

Remember that you can only buy a **whole number** of bags of compost.

3 × 50 litres = 150 litres
4 × 50 litres = 200 litres

You need 168 litres of compost, so you need to buy 4 bags.

Worked example — Target grade 2

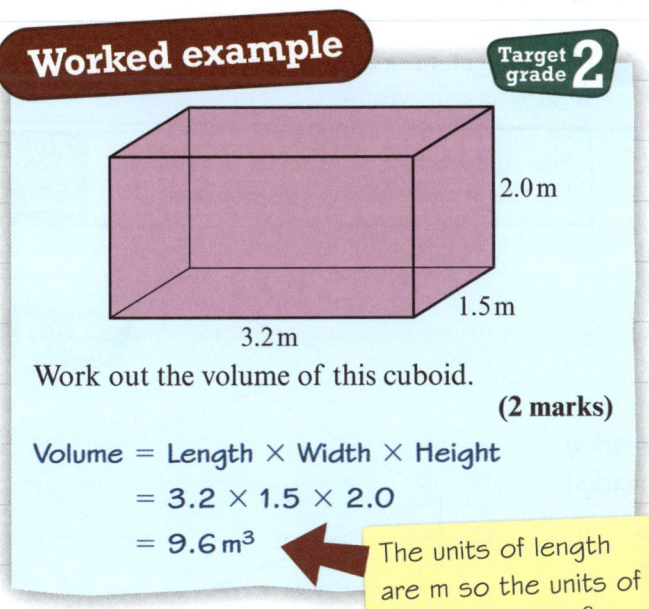

2.0 m, 1.5 m, 3.2 m

Work out the volume of this cuboid. **(2 marks)**

Volume = Length × Width × Height
      = 3.2 × 1.5 × 2.0
      = 9.6 m³

The units of length are m so the units of volume will be m³.

Now try this — Target grade 3

This cuboid has a volume of 390 cm².

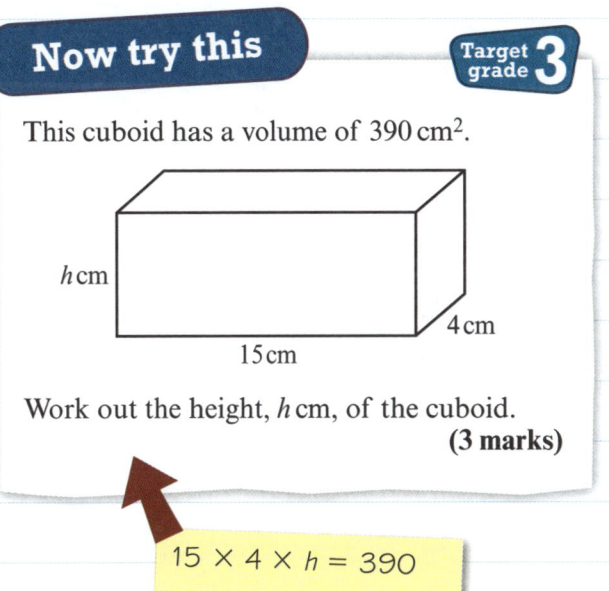

h cm, 15 cm, 4 cm

Work out the height, h cm, of the cuboid. **(3 marks)**

15 × 4 × h = 390

83

Prisms

Volume

A prism is a 3-D solid with a **constant cross-section**. Use this formula to calculate the volume of a prism.

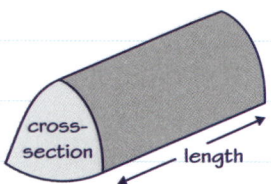

Volume = Area of cross-section × Length

LEARN IT!

Worked example — Target grade 4

The diagram shows a prism. The cross-section is a trapezium. Work out the volume of the prism. **(3 marks)**

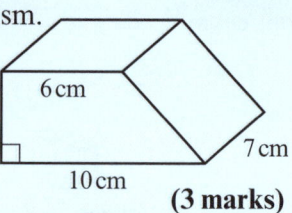

Area of cross-section (trapezium)
$= \frac{1}{2} \times (6 + 10) \times 5 = 40 \text{ cm}^2$

Volume of prism $= 40 \times 7 = 280 \text{ cm}^3$

Surface area

To work out the surface area of a 3-D shape you need to add together the areas of all the faces.

It's a good idea to sketch each face with its dimensions. Remember to include the faces that you can't see.

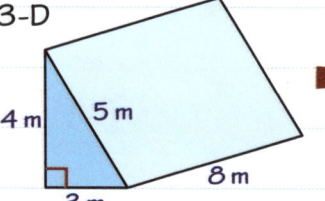

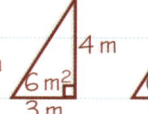

Surface area $= 40 + 32 + 24 + 6 + 6 = 108 \text{ m}^2$

Worked example — Target grade 4

The diagram shows a triangular prism and a cube. They both have the **same** volume. Work out the length of x. **(4 marks)**

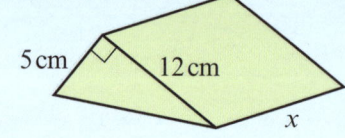 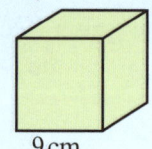

Volume of cube $= 9^3 = 729 \text{ cm}^3$

Volume of prism = Area of cross-section × Length
$= \frac{1}{2} \times 5 \times 12 \times x = 30x$

$30x = 729$

$x = 24.3 \text{ cm}$

Problem solved! Calculate the volume of the cube, and write an expression for the volume of the prism. Set these equal to each other and solve the equation to find x.

Examiners' report

Write down what you are calculating at each stage. You can get marks for your working even if your final answer is wrong.

Real students have struggled with questions like this in recent exams – **be prepared!**

Now try this — Target grade 4

The cross-section of this prism is a right-angled triangle.
(a) Work out the volume of the prism. **(3 marks)**
(b) Work out the total surface area of the prism. **(3 marks)**

Use this formula to work out the areas of the triangular faces:
Area of a triangle $= \frac{1}{2} \times$ Base × Vertical height.

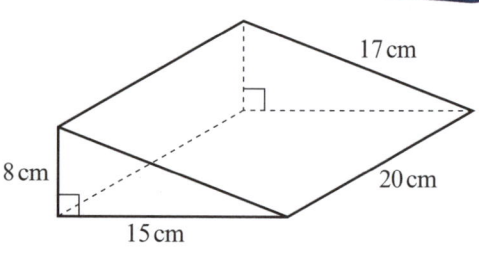

Had a look ☐   Nearly there ☐   Nailed it! ☐

GEOMETRY & MEASURES

Tricky Topic

Units of area and volume

Converting units of area or volume is trickier than converting units of length. You need to remember your area and volume conversions for your exam.

These two squares have the same area.

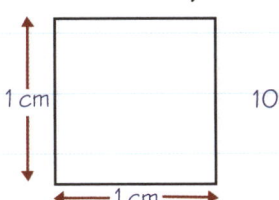
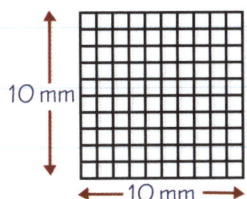

So $1\,cm^2 = 100\,mm^2$.

These two cubes have the same volume.

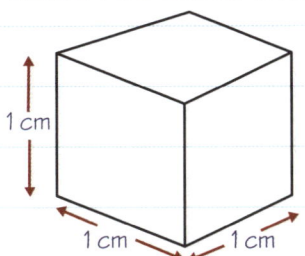
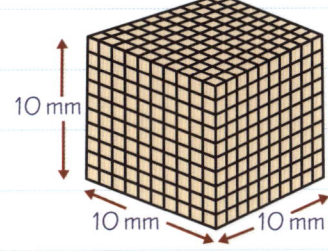

So $1\,cm^3 = 1000\,mm^3$.

Area conversions
$1\,cm^2 = 10^2\,mm^2 = 100\,mm^2$
$1\,m^2 = 100^2\,cm^2 = 10\,000\,cm^2$
$1\,km^2 = 1000^2\,m^2 = 1\,000\,000\,m^2$

Volume conversions
$1\,cm^3 = 10^3\,mm^3 = 1000\,mm^3$
$1\,m^3 = 100^3\,cm^3 = 1\,000\,000\,cm^3$
1 litre $= 1000\,cm^3$
$1\,ml = 1\,cm^3$

There is more on converting metric units on page 61.

Worked example *Target grade 4*

Lead has a density of $11\,350\,kg/m^3$. An antique lead model has a volume of $400\,cm^3$.
Calculate the mass of the model in kg.
(3 marks)

$400 \div 100^3 = 0.0004$
Volume $= 0.0004\,m^3$
Mass $=$ Density $\times$ Volume
$= 11\,350 \times 0.0004$
$= 4.54\,kg$

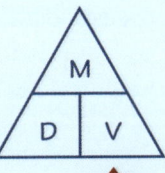

Unit conversion checklist

✓ The multiplier for an area conversion is the length multiplier squared.

✓ The multiplier for a volume conversion is the length multiplier cubed.

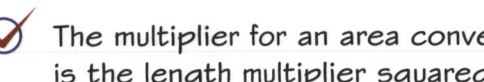
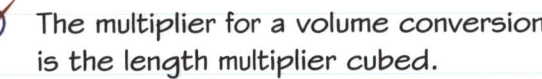
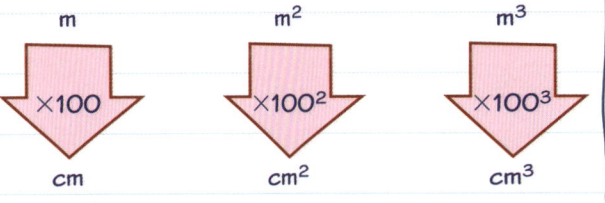

Problem solved! You need to be really careful with the **units** when you are solving any problem involving measures. The units of density are given in kg/m^3, so you need to convert $400\,cm^3$ into m^3 before you calculate. You are converting to a larger unit, so **divide** by 100^3.

For a reminder about density look at page 65.

Now try this

Worked solution video

 **1** Convert
  (a) $2.3\,m^2$ into cm^2 **(1 mark)**
  (b) $400\,mm^3$ into cm^3 **(1 mark)**

 **2** Convert $0.35\,m^3$ to mm^3, giving your answer in standard form. **(2 marks)**

 **3** Jenny applies a force of $600\,N$ to the floor. The total area of her feet is $160\,cm^2$. What is the pressure, in N/m^2, between her and the floor if she stands on both of her feet? **(3 marks)**

Revise pressure on page 66.

85

GEOMETRY & MEASURES — Had a look ☐   Nearly there ☐   Nailed it! ☐

Translations

A translation is a sliding movement. You can describe a translation using a **vector**.

The transformation A → B is a translation by the vector

$$\begin{pmatrix} 4 \\ -3 \end{pmatrix}$$

- The top number describes the horizontal movement:
  - positive number = movement to the right
  - negative number = movement to the left.

The bottom number describes the vertical movement:
- positive number = movement up
- negative number = movement down.

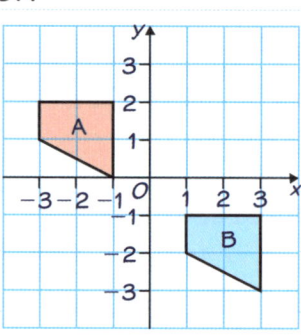

Translated shapes are **congruent**.
For a reminder about congruent shapes have a look at page 109.

Worked example   *Target grade 3*

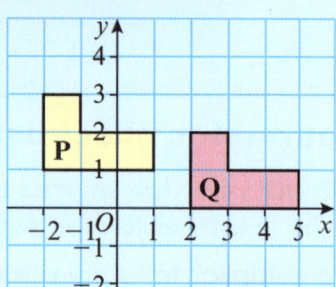

Describe fully the single transformation that will map shape P onto shape Q.   **(2 marks)**

Translation by the vector $\begin{pmatrix} 4 \\ -1 \end{pmatrix}$

To describe a translation fully you need to write the word 'translation' and the vector.

Remember to use **positive** numbers for movement to the right or up, and **negative** numbers for movement to the left or down.

You could also say 'Translation 4 squares to the right and 1 square down'. But don't use the word 'across' to describe a translation as that could mean across in either direction.

For translations, the lengths of sides and the angles in the shapes do not change.

Now try this   *Target grade 3*

Here are some shapes on a grid.
(a) Translate the shaded shape A by the vector $\begin{pmatrix} -5 \\ 4 \end{pmatrix}$

(2 marks)

(b) Write down the vector of the translation that maps shape B onto shape C.   **(2 marks)**

Your vector will be in the form $\begin{pmatrix} x \\ y \end{pmatrix}$ where x is the horizontal movement and y is the vertical movement. Because the shape moves **down** and to the **left** both numbers will be **negative**.

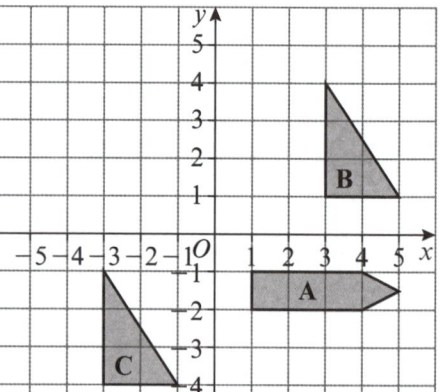

Had a look ☐   Nearly there ☐   Nailed it! ☐

GEOMETRY & MEASURES

Reflections

You can **reflect** a shape in a mirror line. To describe a reflection you need to give the **equation** of the mirror line.

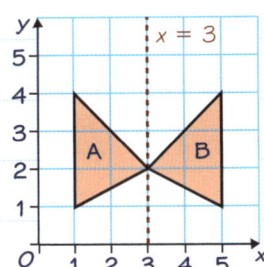

The transformation A → B is a reflection in the line $x = 3$.

Reflected shapes are **congruent**.

Worked example *Target grade 1*

Reflect the shaded shape in the mirror line.

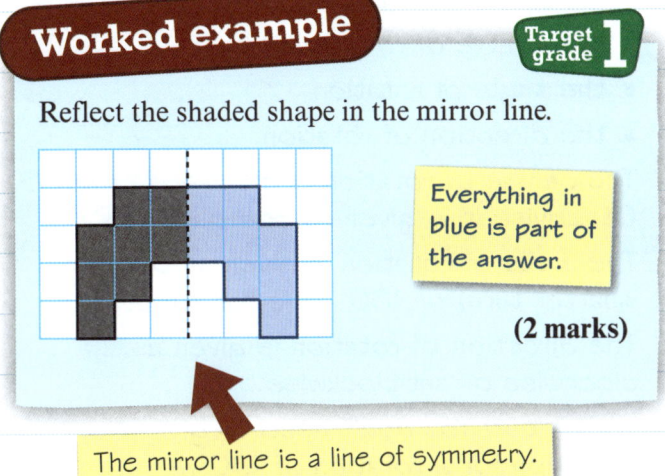

Everything in blue is part of the answer.

(2 marks)

The mirror line is a line of symmetry.

Quick reflections

It is easy to reflect shapes and check your answers using tracing paper.

Trace the original shape including the mirror line.

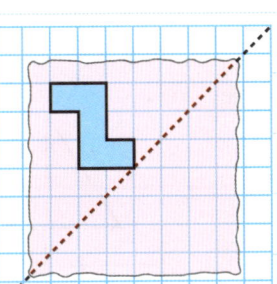

Turn the diagram so that the mirror line is vertical. Turn the tracing paper over, lining up the mirror lines. Trace the shape in the new position.

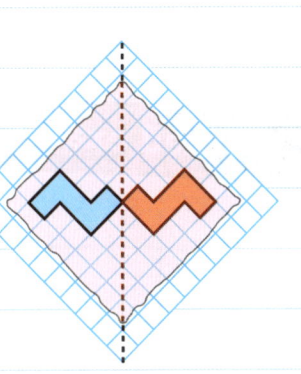

Worked example *Target grade 3*

Describe fully the single transformation that will map shape A onto shape B.

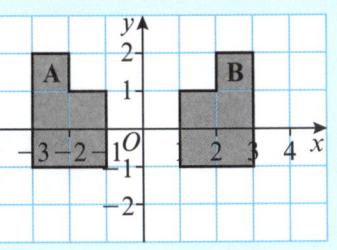

(2 marks)

Reflection in y-axis

Examiners' report

You need to write 'reflection', but you **also** need to describe the mirror line. If it is one of the coordinate axes, it's safer to write this out than to try to give its equation.

Real students have struggled with questions like this in recent exams – **be prepared!**

Now try this *Target grade 4*

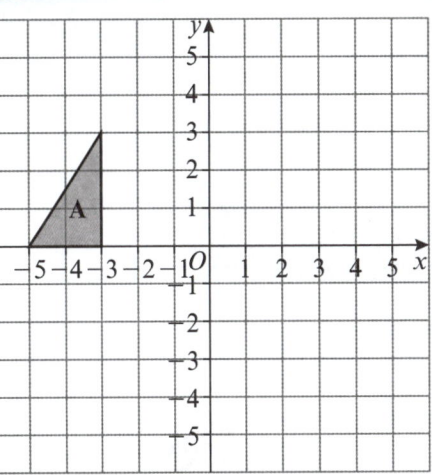

(a) Reflect triangle A in the y-axis. Label the image B. **(1 mark)**
(b) Reflect triangle B in the x-axis. Label the image C. **(1 mark)**
(c) Reflect triangle C in the line $y = x$. Label the image D. **(2 marks)**
(d) Describe fully the single transformation that maps triangle D onto triangle A. **(2 marks)**

GEOMETRY & MEASURES Had a look ☐ Nearly there ☐ Nailed it! ☐

Rotations

To describe a **rotation** you need to give
- the centre of rotation
- the angle of rotation
- the direction of rotation.

The centre of rotation is often the origin O. Otherwise it is given as coordinates.

The angle of rotation is given as 90° (one quarter turn) or 180° (one half turn).

The direction of rotation is given as clockwise or anticlockwise.

Watch out! You don't need to give a direction for a rotation of 180°.

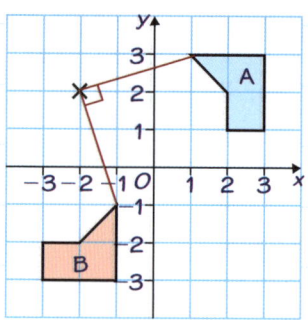

A to B: Rotation 90° clockwise about the point (−2, 2).

Give the angle in degrees. Don't say 'a quarter turn'.

You are allowed to ask for tracing paper in the exam. This makes it really easy to rotate shapes and check your answers.

Rotated shapes are **congruent**.

For a reminder about congruent shapes have a look at page 109.

Worked example Target grade 4

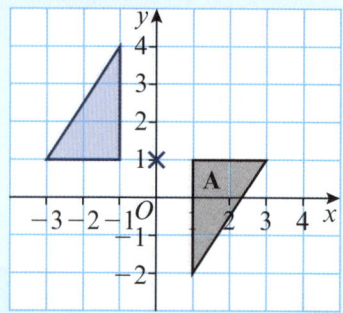

Everything in blue is part of the answer.

On the grid, rotate shape A 180° about (0, 1). **(2 marks)**

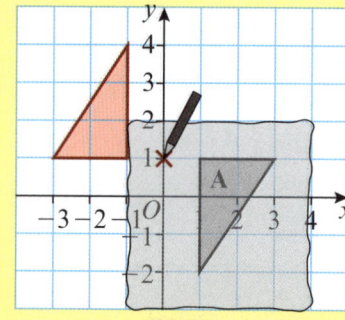

Mark the centre of rotation (0, 1) with a X.

Trace the shape and put your pencil or compass on the X. Rotate the tracing paper to rotate the shape.

Now try this Target grade 4

Describe fully the rotation that maps shape P onto
(a) shape Q **(3 marks)**
(b) shape R. **(3 marks)**

You need to write the word 'rotation' and give
- the angle of the turn
- the direction of the turn
- the centre of rotation.

Worked solution video

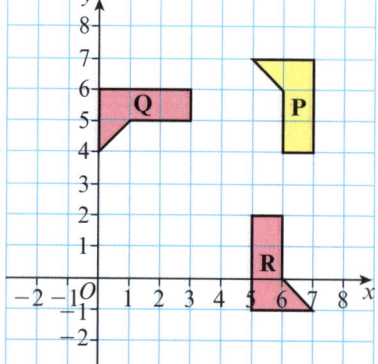

Had a look ☐ Nearly there ☐ Nailed it! ☐

GEOMETRY & MEASURES

Enlargements

To describe an enlargement you need to give the scale factor and the centre of enlargement.

The **scale factor** of an enlargement tells you how much each length is multiplied by.

$$\text{Scale factor} = \frac{\text{Enlarged length}}{\text{Original length}}$$

Lines drawn through corresponding points on the object (A) and image (B) meet at the **centre of enlargement**.

When the scale factor is between 0 and 1, image B is **smaller** than object A.

For enlargements, angles in shapes do not change but lengths of sides do change.

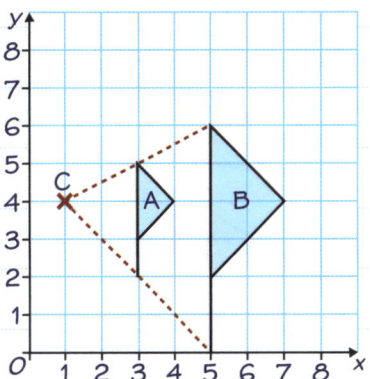

A to B: each point on B is twice as far from C as the corresponding point on A.
Enlargement with scale factor 2, centre (1, 4)

Worked example *Target grade 5*

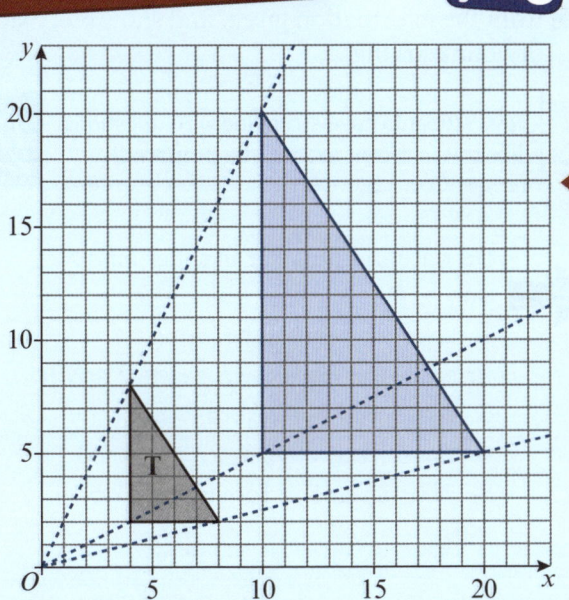

On the grid, enlarge triangle T with a scale factor of $2\frac{1}{2}$ and centre (0, 0) **(3 marks)**

1. Draw lines from the centre of enlargement through each vertex of the triangle.

2. For each vertex, multiply the vertical and horizontal distances from the centre of enlargement by $2\frac{1}{2}$

   For the top vertex:
   Horizontal distance = $4 \times 2\frac{1}{2} = 10$
   Vertical distance = $8 \times 2\frac{1}{2} = 20$

   The corresponding vertex on the image is 10 squares horizontally and 20 squares vertically from the centre of enlargement.

3. Join up your vertices with straight lines.

Check it!

Each length on the image should be $2\frac{1}{2}$ times the corresponding length on the object. The image is **mathematically similar** to the object, so check it looks the same shape.

There is more on similar shapes on pages 109 and 110.

Now try this

 Triangle A is shown on the grid.
(a) Enlarge triangle A with a scale factor of 2 and centre of enlargement (6, 5). Label the image B. **(2 marks)**

 (b) Enlarge triangle A with a scale factor of $\frac{1}{2}$ and centre of enlargement (−7, 3). Label the image C. **(3 marks)**

The scale factor is a **fraction** between 0 and 1 so the image will be **smaller** than the object.

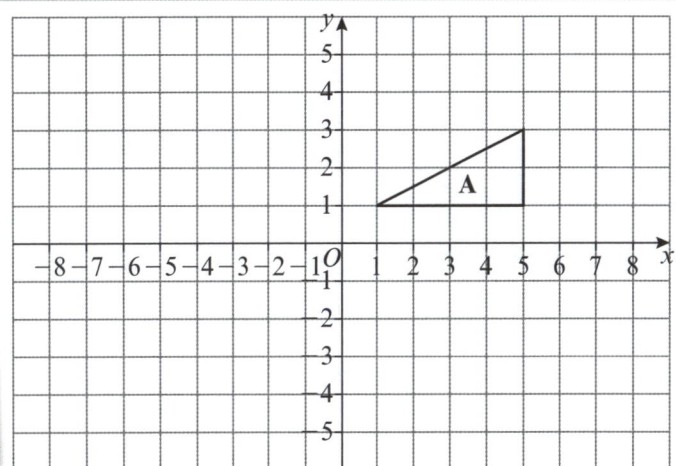

89

GEOMETRY & MEASURES — Had a look ☐ Nearly there ☐ Nailed it! ☐

Pythagoras' theorem

Pythagoras' theorem is a really useful rule. You can use it to find the length of a missing side in a right-angled triangle.

LEARN IT!

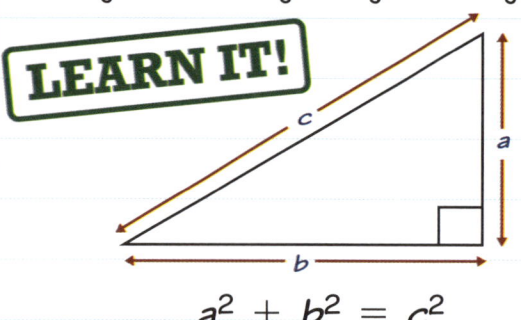

$a^2 + b^2 = c^2$

Pythagoras checklist
- ✓ short^2 + short^2 = long^2
- ✓ Right-angled triangle.
- ✓ Lengths of two sides known.
- ✓ Length of third side missing.
- ✓ Learn this. It won't be given in the exam.

Worked example (Target grade 4)

This right-angled triangle has sides x, 17 cm and 8 cm.

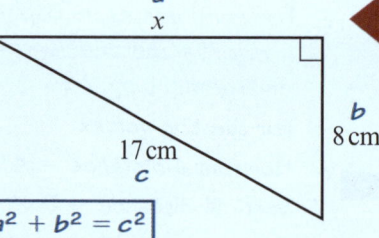

$a^2 + b^2 = c^2$

Show that $x = 15$ cm. (2 marks)

$x^2 + 8^2 = 17^2$
$x^2 = 17^2 - 8^2$
$ = 289 - 64$
$ = 225$
$x = \sqrt{225} = 15$ cm

Everything in blue is part of the answer.

Examiners' report

This question asks you to **show that** $x = 15$. The safest way to do this is to find the value of x from the information given, and show **every step** of your working.

Real students have struggled with questions like this in recent exams – **be prepared!**

Be careful when the missing length is one of the **shorter** sides.
1. Label the longest side c.
2. Label the other two sides a and b.
3. Write out Pythagoras' theorem.
4. Substitute any values you know.
5. Rearrange the formula and solve.

Pythagoras questions come in lots of different forms. Just look for the right-angled triangle.

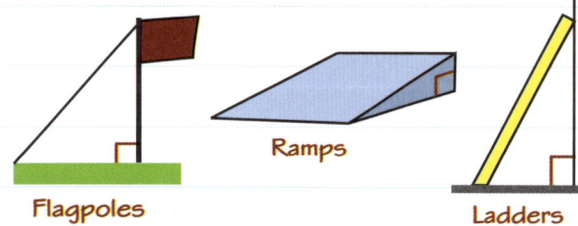

Flagpoles   Ramps   Ladders

Calculator skills

Use these buttons to find squares and square roots with your calculator.

You might need to use the [S⇔D] key to get your answer as a decimal number.

Now try this (Target grade 4)

(a) Work out the value of y. **(2 marks)**
(b) Use your value of y to work out the value of z. **(2 marks)**

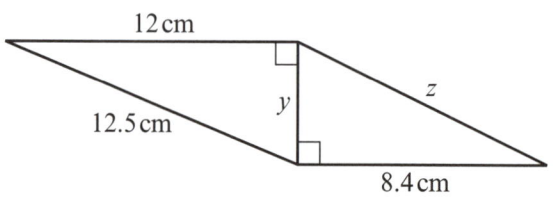

90

Had a look ☐   Nearly there ☐   Nailed it! ☐

GEOMETRY & MEASURES

Tricky Topic

Line segments

A line segment is a section of a straight line between two points. You can use Pythagoras' theorem to find the length of a line segment.

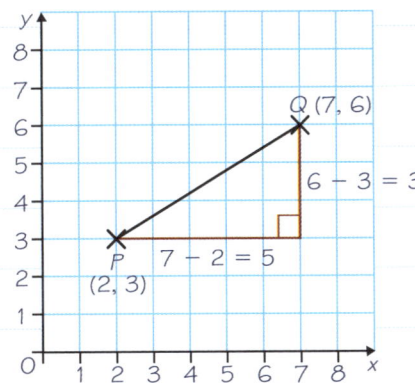

Draw a right-angled triangle with PQ as its longest side.

$PQ^2 = 5^2 + 3^2$
$= 25 + 9$
$= 34$
$PQ = \sqrt{34}$
$= 5.8309...$
$= 5.83$ (2 decimal places)

For a reminder of how to find the mid-point of a line segment have a look at page 36.

Worked example

Target grade 4

Point A has coordinates (2, 5)
Point B has coordinates (3, −2)
Calculate the length of the line segment AB.
Give your answer correct to 2 decimal places. **(3 marks)**

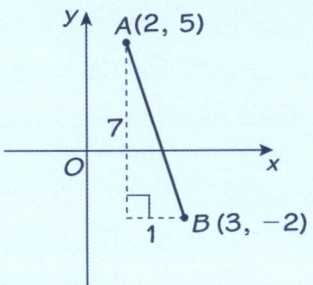

Horizontal distance = 3 − 2 = 1
Vertical distance = 5 − −2 = 7
$AB^2 = 1^2 + 7^2$
$= 50$
$AB = \sqrt{50}$
$= 7.07106... = 7.07$ (2 d.p.)

1. Sketch x- and y-axes.
2. Mark points A and B on your sketch.
3. Draw a right-angled triangle with AB as its hypotenuse.
4. Work out the length of the two short sides of the triangle.
5. Use Pythagoras' theorem to work out the length of AB.
6. Round your answer to 2 decimal places.

Watch out!

Be really careful if any of the coordinates is negative.

The height of the triangle is
5 − −2 = 5 + 2 = 7

You are working out a length so you can only substitute positive numbers into Pythagoras' theorem.

Now try this

Target grade 4

1
P (−3, 5)
Q (4, −1)

Work out the length of the line segment PQ. Give your answer correct to 2 decimal places. **(3 marks)**

Worked solution video

2 Find the perimeter of the isosceles triangle ABC. Give your answer to 2 decimal places.

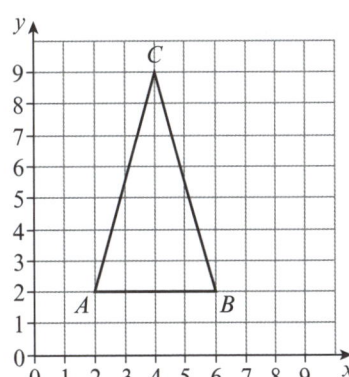

(4 marks)

GEOMETRY & MEASURES

Had a look ☐   Nearly there ☐   Nailed it! ☐

Trigonometry 1

You can use the trigonometric ratios to find the size of an angle in a right-angled triangle. You need to know the lengths of two sides of the triangle.

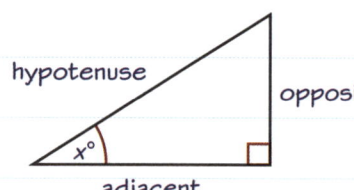

The sides of the triangle are labelled relative to the **angle** you need to find.

Trigonometric ratios — **LEARN IT!**

$\sin x° = \dfrac{\text{opp}}{\text{hyp}}$ (remember this as S^O_H)

$\cos x° = \dfrac{\text{adj}}{\text{hyp}}$ (remember this as C^A_H)

$\tan x° = \dfrac{\text{opp}}{\text{adj}}$ (remember this as T^O_A)

You can use $S^O_H C^A_H T^O_A$ to remember these rules for trig ratios.
These rules only work for **right-angled** triangles.

Worked example — Target grade 5

Calculate the size of angle x. **(3 marks)**

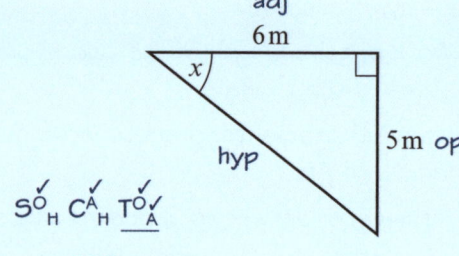

$S^O_H \checkmark \; C^A_H \checkmark \; T^O_A \checkmark$

$\tan x° = \dfrac{\text{opp}}{\text{adj}} = \dfrac{5}{6}$

$x° = 39.80557109° = 39.8°$ (to 3 s.f.)

Label the **hyp**otenuse first — it's the longest side.
Then label the side **adj**acent to the angle you want to work out.
Finally label the side **opp**osite the angle you want to work out.
Remember $S^O_H C^A_H T^O_A$. You know **opp** and **adj** here so use T^O_A.
Do **not** 'divide by tan' to get x on its own. You need to use the $\tan^{-1}$ function on your calculator.

$\tan^{-1}\left(\dfrac{5}{6}\right)$
39.80557109

Write all the figures on your calculator display and then round your answer.

Using your calculator

To find a missing angle using trigonometry you have to use one of these functions.
$$\sin^{-1} \quad \cos^{-1} \quad \tan^{-1}$$
These are called **inverse trigonometric** functions. They are the inverse operations of sin, cos and tan.
Make sure that your calculator is in degree mode. Look for the D symbol at the top of the display.

Now try this — Target grade 5

Work out the size of angle x in each of these triangles. Give your answers correct to 1 decimal place.

(a) 4.3 cm, 6.1 cm   **(3 marks)**

(b) 11.2 cm, 7.5 cm   **(3 marks)**

(c) 84 mm, 127 mm   **(3 marks)**

Had a look ☐ Nearly there ☐ Nailed it! ☐

GEOMETRY & MEASURES

Tricky Topic

Trigonometry 2

You can use the trigonometric ratios to find the length of a missing side in a right-angled triangle. You need to know the length of another side and the size of one of the acute angles.

Worked example

Target grade 5

Calculate the length of side a. **(3 marks)**

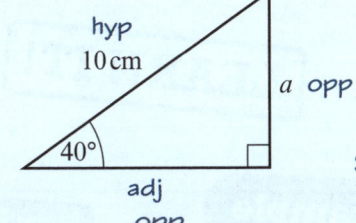

$S\overset{\checkmark}{O}_H C\overset{\checkmark}{A}_H T\overset{\checkmark}{O}_A$

$\sin x° = \dfrac{opp}{hyp}$

$\sin 40° = \dfrac{a}{10}$

$a = 10 \times \sin 40°$

$= 6.42787...$

$= 6.43\,cm$ (to 3 s.f.)

Label the sides of the triangle relative to the 40° angle. Write $S^O{}_H C^A{}_H T^O{}_A$ and tick the pieces of information you have. You need to use $S^O{}_H$ here.

Write the values you know in the rule and replace **opp** with a. You can solve this equation to find the value of a.

Write down at least four figures from the calculator display before giving your final answer correct to 3 significant figures.

Check it!
Side a must be shorter than the hypotenuse. 6.43 cm looks about right. ✓

Angles of elevation and depression

Some trigonometry questions will involve angles of elevation and depression.

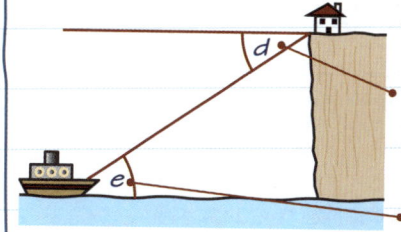

The angle of depression of the ship from the house.

The angle of elevation of the house from the ship.

Angles of elevation and depression are always measured from the horizontal.

In this diagram, $d = e$ because they are alternate angles.

Now try this

Target grade 5

In part (c) a is the hypotenuse. It will be on the bottom of the fraction when you substitute, so be careful with your calculation.

Work out the length of side a in each of these triangles. Give your answers correct to 1 decimal place.

(a) 5.6 cm, 38°, side a

(b) a, 73°, 18.5 cm

(c) a, 52°, 12.4 cm

(3 marks) (3 marks) (3 marks)

Exact trigonometry values

You might need to answer questions involving sin, cos and tan **without a calculator**. Make sure you are confident with pages 92 and 93 before having a look at this page.

1 This triangle shows you the values of sin, cos and tan for 30° and 60°.

$\sin 30° = \dfrac{1}{2}$ $\sin 60° = \dfrac{\sqrt{3}}{2}$

$\cos 30° = \dfrac{\sqrt{3}}{2}$ $\cos 60° = \dfrac{1}{2}$

$\tan 30° = \dfrac{1}{\sqrt{3}}$ $\tan 60° = \sqrt{3}$

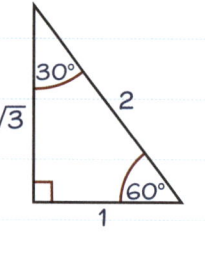

2 This triangle shows you the values of sin, cos and tan for 45°.

$\sin 45° = \dfrac{1}{\sqrt{2}}$

$\cos 45° = \dfrac{1}{\sqrt{2}}$

$\tan 45° = 1$

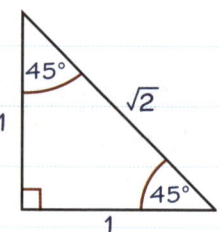

LEARN IT!

3 You also need to know the values for 0° and 90°.

$\sin 0° = 0$ $\sin 90° = 1$

$\cos 0° = 1$ $\cos 90° = 0$

$\tan 0° = 0$

tan 90° is **undefined**. If you enter it into a calculator you get an error.

You might get a question like this on your **non-calculator** paper, so make sure you learn the values of cos, sin and tan given above. If a triangle is **described** like this make sure you **sketch** it before doing any working.

Worked example *Target grade 5*

Triangle ABC has a right angle at B. Angle $BAC = 60°$. $AC = 14$ cm. Calculate the length of AB. **(3 marks)**

$\cos x° = \dfrac{\text{adj}}{\text{hyp}}$

$\cos 60° = \dfrac{AB}{14}$

$AB = 14 \times \cos 60°$

$= 14 \times \dfrac{1}{2} = 7$ cm

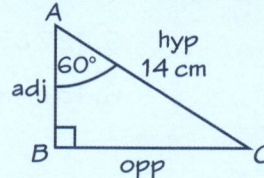

Worked example *Target grade 5*

Amy arranges identical triangles like this around a point to make a pattern. Show that she can fit exactly 12 triangles around a point. **(4 marks)**

$\sin x° = \dfrac{\text{opp}}{\text{hyp}} = \dfrac{3}{6} = \dfrac{1}{2}$

So $x° = 30°$

Everything in blue is part of the answer.

There are 360° around a point and 30 × 12 = 360 so she can fit exactly 12 triangles around a point.

Problem solved! This is a non-calculator question, so you will need to use your knowledge of exact trigonometric values. Make sure you **write a conclusion** explaining how your working shows the answer.

You need to be able to answer this question **without** a calculator.

Now try this *Target grade 5*

A vertical flagpole AB is supported with a wire AC at an angle of 60° to the ground. The base of the wire is 2.4 m from the base of the flagpole. Show that the length of the wire AC is 4.8 m. **(3 marks)**

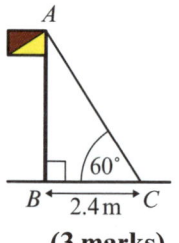

Had a look ☐   Nearly there ☐   Nailed it! ☐   **GEOMETRY & MEASURES**

Measuring and drawing angles

You should only measure angles in your exam if you are told that a diagram is drawn accurately.

Measuring angles

1 A protractor measures angles in degrees.

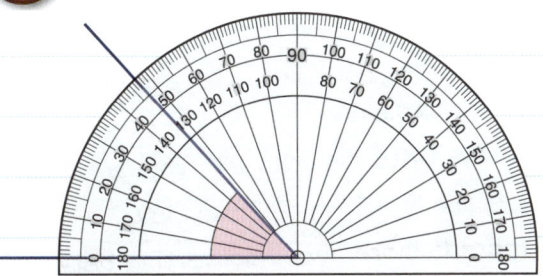

Use the scale that starts with 0 on one of the lines of the angle.
Here, use the outside scale.

Place the centre of the protractor on the point of the angle.
Line up the zero line with one line of the angle.
Read the size of the angle off the scale.
This angle is 47°.

2

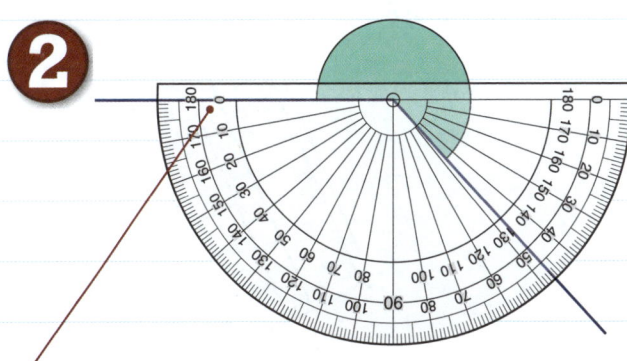

Use the scale that starts with 0 on one of the lines of the angle.
Here, use the inside scale.

To measure an angle bigger than 180° measure the smaller angle then subtract the answer from 360°.
360 − 133 = 227
The marked angle is 227°.

> Estimate the size of an angle before measuring it. This lets you check that your answer is sensible.

Drawing angles

1 Draw an angle of 23°.

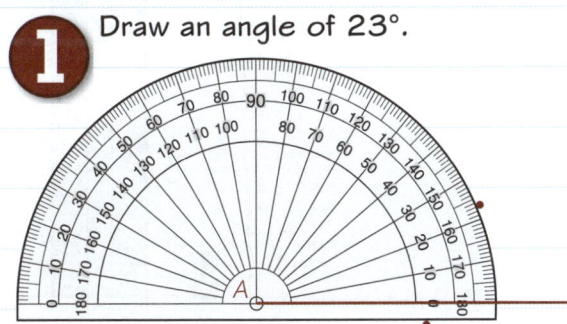

Use the scale that starts with 0 on one of the lines of the angle.
Here, use the inside scale.

Use a ruler to draw one line of the angle, AB.

Place the centre of your protractor on one end of the line. The zero line needs to lie along your line.

Find 23° on the scale.
Draw a dot to mark this point.

2

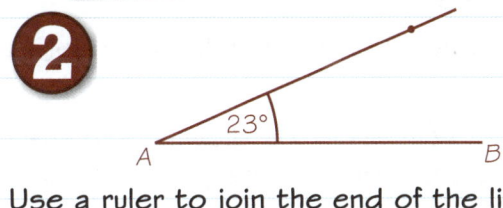

Use a ruler to join the end of the line and your dot with a straight line.

Draw in the angle curve and label your angle 23°.

Now try this *Target grade 1*

1 Measure the size of each of these angles.
  (a)   (b)

Extend the lines with a pencil and ruler.

          (1 mark)   (1 mark)

2 Use a protractor to draw an angle of 124°.
          (1 mark)

GEOMETRY & MEASURES | Had a look ☐ | Nearly there ☐ | Nailed it! ☐

Measuring lines

You need to be able to use a ruler to draw and measure straight lines accurately.
Don't measure lines in your exam unless the question tells you that the diagram is accurate.

Worked example
Target grade 1

Here is a line *AB*.

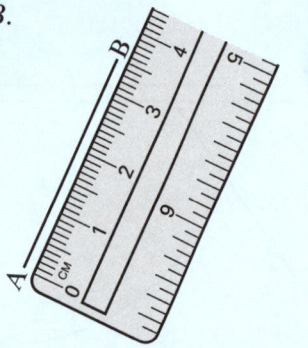

Measure the length of the line *AB*. **(1 mark)**

35 mm

> Line up the 0 mark on your ruler carefully with the start of the line at A.
>
> Always measure to the nearest mm.
>
> Make sure your ruler doesn't move while you're measuring the line.
>
> Always write the units with your answer.
>
> This line is 35 mm or 3.5 cm long.
>
> There is more about converting between cm and mm on **page 61**.

Drawing lines checklist

✓ Check whether you are working in cm or mm.
✓ Start the line at the 0 mark on your ruler.
✓ Hold your ruler firmly.
✓ Use a sharp pencil.
✓ Draw to the nearest mm.
✓ Label the length you have drawn.

Estimating

You can use lengths that you know to estimate other lengths.
This diagram shows a man standing at the bottom of a cliff.
The man is 3 cm tall in the drawing and the cliff is 12 cm tall.
This means the cliff is 4 times as tall as the man.
A good estimate for the height of an adult male is 1.8 m.
$4 \times 1.8 = 7.2$
This means a good estimate for the height of the cliff is 7.2 m.

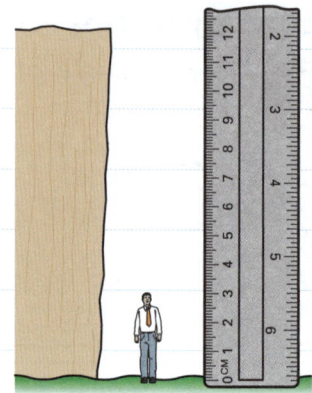

Now try this
Target grade 1

1. Measure the lengths of these lines.
   (a) _____
   (b) _____
   (c) _____
   (d) _____ **(4 marks)**

   Measure to the nearest mm.

2. The picture of a surfboard and a London Bus have been drawn accurately to the same scale. The real length of the surfboard is 5 feet. Estimate the real length of the London Bus. **(3 marks)**

Had a look ☐   Nearly there ☐   Nailed it! ☐

GEOMETRY & MEASURES

Hot Topic

Plans and elevations

Plans and elevations are 2-D drawings of 3-D shapes as seen from different directions.

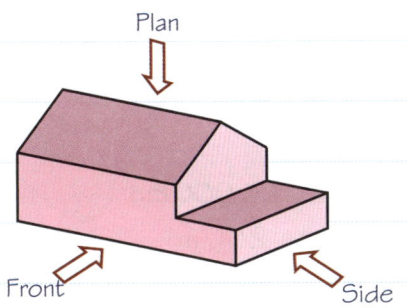

A **sketch** of the shape would look like this.

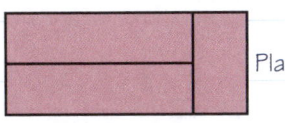

The **plan** is the view from above.

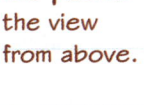

This line shows a change in depth.

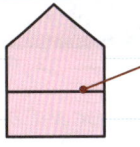

The **front elevation** is the view from the front.

The **side elevation** is the view from the side.

Nets

A **net** of a 3-D shape is a flat pattern that can be folded up to make the shape. This is a net of a cube.

To decide if a pattern is a net of a 3-D shape imagine trying to fold it up to make the shape.

Worked example   Target grade 3

The diagram shows a solid shape.

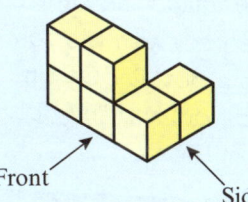

On the grid below draw a plan, front and side elevations of the shape.

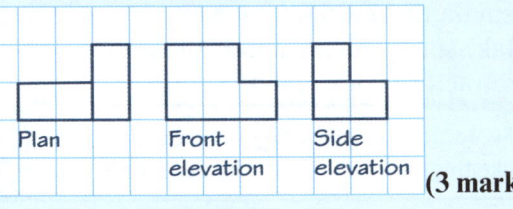

(3 marks)

Imagine tracing an image of the shape on each side of a box.
Unfold the box to get your plan and elevations.

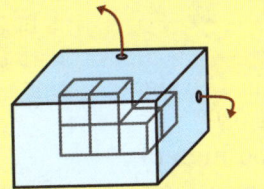

Put lines within the plan and side elevation to show where there is a change in height or depth.

Draw lines on your plan and elevations to show where the height of the shape changes.

Now try this   Target grade 3

This solid is made from centimetre cubes.
(a) On separate 1 cm grids draw
 (i) the plan view of the solid **(1 mark)**
 (ii) the front elevation of the solid **(1 mark)**
 (iii) the side elevation of the solid. **(1 mark)**
(b) Work out the total surface area of the solid.
(2 marks)

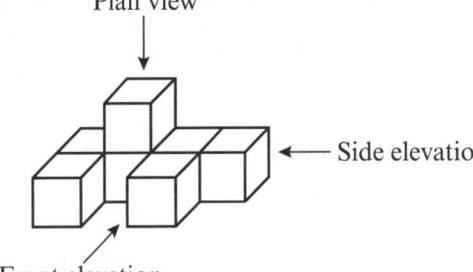

97

Scale drawings and maps

This is a **scale drawing** of the Queen Mary II cruise ship.

Scale = 1 : 1000

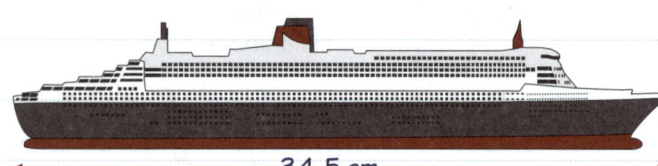

←— 34.5 cm —→

You can use the scale to work out the length of the actual ship.

34.5 × 1000 = 34 500

The ship is 34 500 cm or 345 m long.

Map scales

Map scales can be written in different ways:
- ✓ 1 to 25 000
- ✓ 1 cm represents 25 000 cm
- ✓ 1 cm represents 250 m
- ✓ 4 cm represent 1 km

Worked example *Target grade 3*

The diagram shows a scale drawing of a port and a lighthouse. This diagram is drawn accurately.

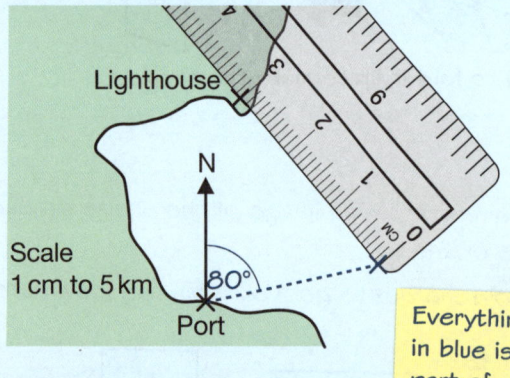

Everything in blue is part of the answer.

A boat sails 12 km in a straight line on a bearing of 080°.
(a) Mark the new position of the boat with a cross. **(2 marks)**
(b) How far away is the boat from the lighthouse? Give your answer in km. **(1 mark)**

15 km

You can revise bearings on page 102. Start by working out how far the boat is from the port on the scale drawing.

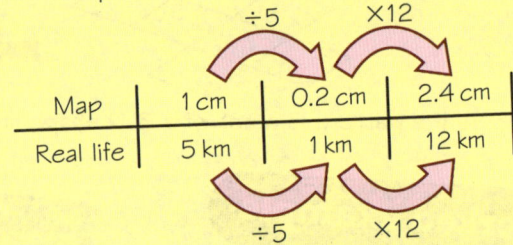

Map	1 cm	0.2 cm	2.4 cm
Real life	5 km	1 km	12 km

Now place the centre of your protractor on the port with the zero line pointing north. Put a dot at 80°. Line up your ruler between the port and the dot. Draw a cross 2.4 cm from the port.

Examiners' report

3 cm on the drawing represents 15 km in real life. Make sure you have a millimetre ruler **and** a protractor with you in the exam.

Real students have struggled with questions like this in recent exams – **be prepared!**

Now try this

Worked solution video

Read the whole question before you start. You need to give your answer in cm, so you could start by converting 20 m into cm.

Target grade 3

The Angel of the North is a 20 m tall statue. Penny makes a model of the statue using a scale of 1 : 25

What is the height of the model of the statue? Give your answer in centimetres. **(2 marks)**

Had a look ☐ Nearly there ☐ Nailed it! ☐

GEOMETRY & MEASURES

Constructions 1

You might be asked to construct a perpendicular line in any of these three ways.

Worked example
Target grade 4

Use ruler and compasses to **construct** the perpendicular to the line segment AB that passes through point P.

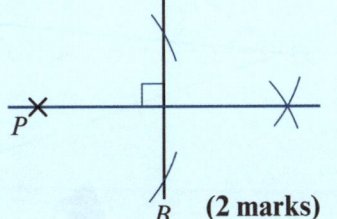

Everything in blue is part of the answer.

(2 marks)

Use your compasses to mark two points on the line an equal distance from P. Keep the compasses the same and draw two arcs with their centres at these points.

Worked example
Target grade 4

Use ruler and compasses to **construct** the perpendicular to the line segment AB that passes through point P.

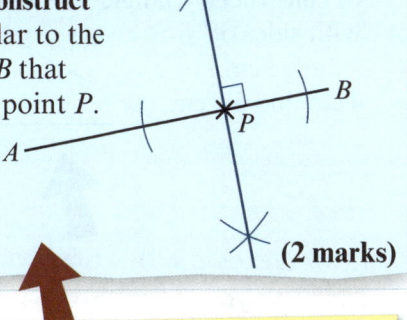

(2 marks)

Use your compasses to mark two points an equal distance from P. Then widen your compasses and draw arcs with their centres at these two points.

Worked example
Target grade 4

Use ruler and compasses to **construct** the perpendicular bisector of the line AB.

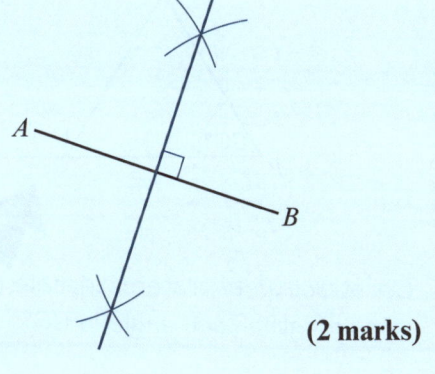

(2 marks)

Examiners' report

Use your compasses to draw intersecting arcs with centres at A and B. Remember that to get full marks you have to be **accurate** and show **all** your construction lines.

Real students have struggled with questions like this in recent exams – **be prepared!**

Constructions checklist
- ✓ Use good compasses with stiff arms.
- ✓ Use a sharp pencil.
- ✓ Use a transparent ruler.
- ✓ Mark any angles.
- ✓ Label any lengths.
- ✓ Show all construction lines.

Now try this
Target grade 4

Construct the perpendicular bisector of the line AB. (2 marks)

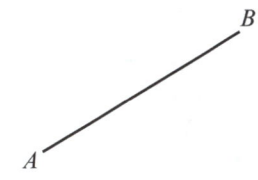

Remember to show **all** your construction marks.

Worked solution video

99

GEOMETRY & MEASURES

Had a look ☐ Nearly there ☐ Nailed it! ☐

Constructions 2

You need to know all of these constructions for your exam.

Worked example *Target grade 3*

Use ruler and compasses to **construct** a triangle with sides of length 3 cm, 4 cm and 5.5 cm.

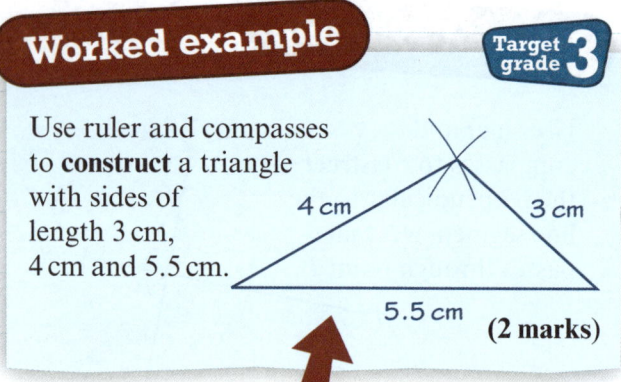

(2 marks)

Draw one side with a ruler and label it. Then use your compasses to find the other vertex.

Worked example *Target grade 4*

Use ruler and compasses to **construct** a 45° angle at P.

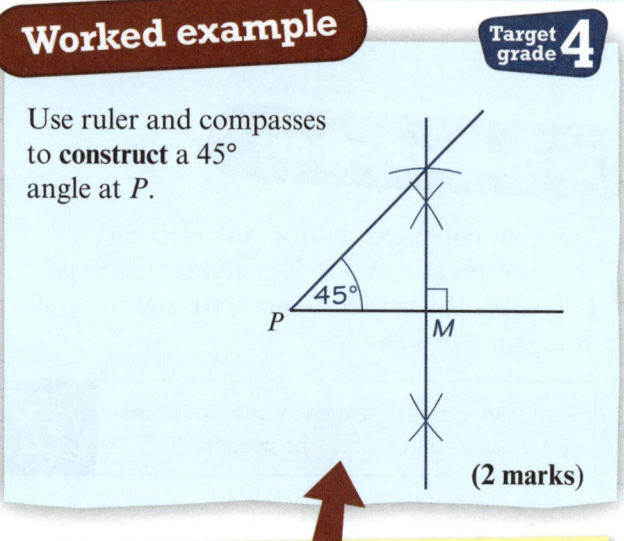

(2 marks)

Construct the perpendicular bisector of the line. Mark the mid-point M. Now set your compasses to the distance PM. Draw an arc on your bisector and join this point to P with a ruler.

Worked example *Target grade 4*

Use ruler and compasses to **construct** the bisector of angle PQR.

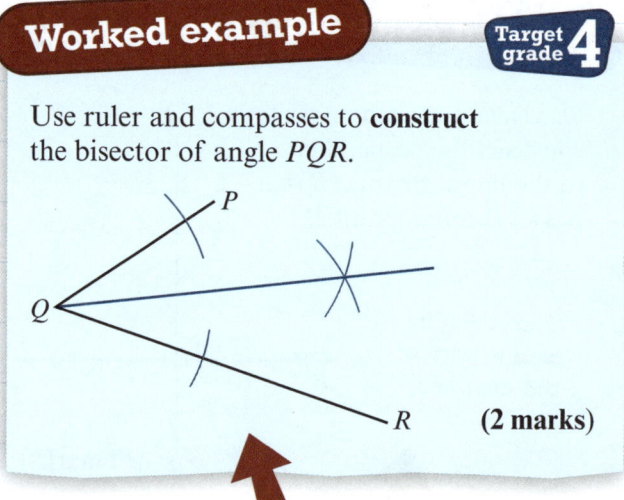

(2 marks)

Mark points on each arm an equal distance from Q. Then use arcs to find a third point an equal distance from these two points.

Worked example *Target grade 4*

Use ruler and compasses to **construct** a 60° angle at P.

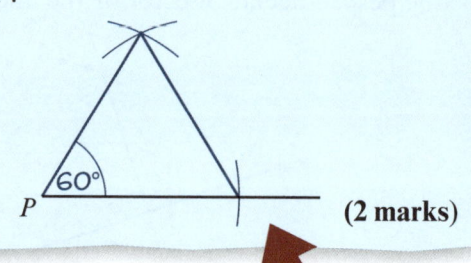

(2 marks)

Construct an equilateral triangle (all sides the same length). Each angle is 60°.

Now try this *Target grade 4*

1 Construct the bisector of the angle ABC.
(2 marks)

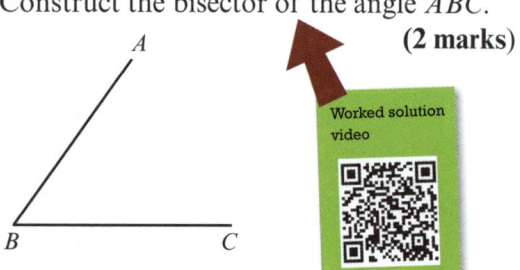

2 Use ruler and compasses to construct an angle of 30°.
(3 marks)

Do this on a separate piece of paper. You can construct a 30° angle by constructing a 60° angle and then bisecting it.

100

Had a look ☐ **Nearly there** ☐ **Nailed it!** ☐

GEOMETRY & MEASURES

Loci

A **locus** is a set of points that satisfy a condition. You can construct loci using ruler and compasses. A set of points can lie inside a **region** rather than on a line or curve.

The locus of points which are 7 cm from A is the circle, centre A.

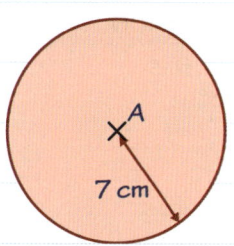

The region of points less than 7 cm from A lies inside this circle.

The locus of points which are the **same distance** from B as from C is the perpendicular bisector of BC.

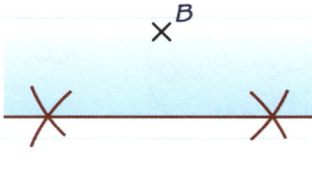

Points in the shaded region are closer to B than to C.

The locus of points which are 2 cm away from ST consists of two semicircles and two straight lines.

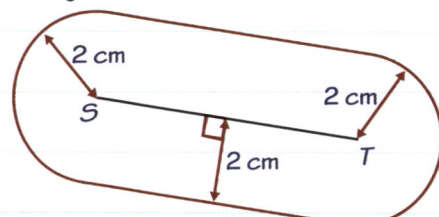

The shortest distance between a point and a line is the **perpendicular** from the point to the line.

Combining conditions

You can be asked to shade a region which satisfies more than one condition.

Here, the shaded region is more than 6 cm from point D **and** closer to line BC than to line AD.

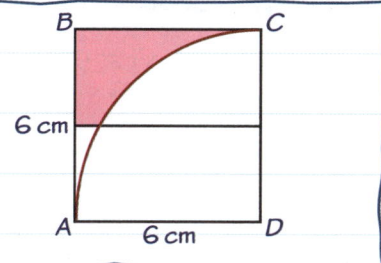

Worked example

Target grade 4

This diagram is drawn accurately and shows part of a beach and the sea.
1 cm represents 20 m.
There is a lifeguard tower at point P. Public swimming is allowed in a region of the sea less than 30 m from the lifeguard tower.
Shade this region on the diagram. **(2 marks)**

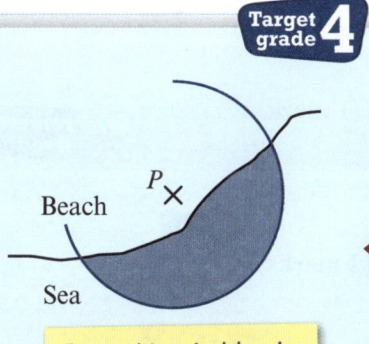

Everything in blue is part of the answer.

1 cm represents 20 m so 1.5 cm represents 30 m.
There is more about scale drawing on **page 98**.
Set your compasses to 1.5 cm.
Set your compasses accurately by placing the point **on top** of your ruler at the 0 mark.

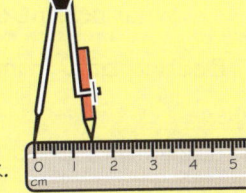

Now try this

Target grade 4

A piece of card in the shape of an equilateral triangle ABC is placed on a horizontal straight line.

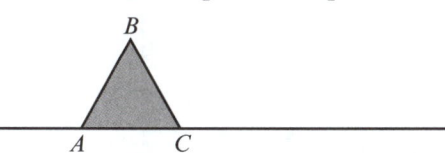

The card is first rotated 120° clockwise about C.
The card is then rotated 120° clockwise about B.
Draw the locus of the vertex A. **(3 marks)**

Start with your compass point at C and your pencil at A and draw an arc. You need to work out where B touches the ground and move your compass point to there.

101

GEOMETRY & MEASURES

Had a look ☐   Nearly there ☐   Nailed it! ☐

Bearings

Bearings are measured **clockwise** from **north**.

Bearings always have **three figures**. You need to add zeros if the angle is less than 100°. For instance, in this diagram the bearing of B from A is 048°.

You can measure a bearing bigger than 180° by measuring this angle and subtracting it from 360°.

The bearing of C from A is 360° − 109° = 251°

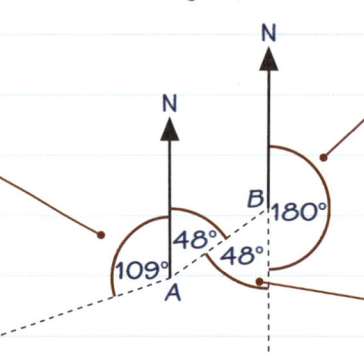

You can work out a reverse bearing by adding or subtracting 180°.

The bearing of A from B is 180° + 048° = 228°

These are alternate angles.

Worked example

Target grade 3

Three paths meet at O. B is due east of O.

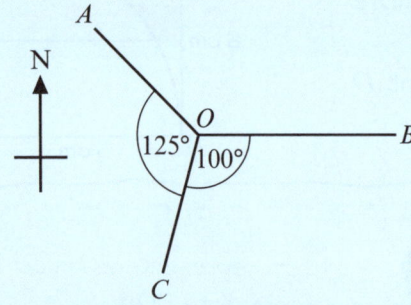

(a) Jake walks from O to A.
What bearing does he walk on? **(1 mark)**

90° + 100° + 125° = 315°

(b) Delvinder walks from C to O.
What bearing does she walk on? **(2 marks)**

Bearing of C from O = 90° + 100°
= 190°

Bearing of O from C = 190° − 180° = 010°

Compass points

You need to know the compass points:

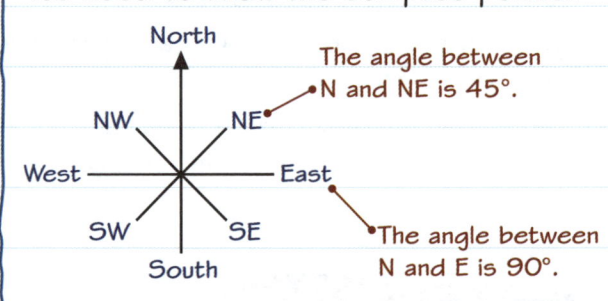

The angle between N and NE is 45°.

The angle between N and E is 90°.

Problem solved! You can work out a **reverse** bearing by adding or subtracting 180°. If the angle is greater than 180° then subtract. Remember to write your final answer as a **three-figure bearing**. You need to write 010°, not just 10°.

Now try this

Target grade 3

An aircraft flies from A to B. The grid shows the positions of A and B.
(a) Use the grid to work out the actual distance AB. **(1 mark)**
(b) Measure and write down the three-figure bearing of B from A. **(1 mark)**
(c) The aircraft then flies to C. The bearing of C from A is 120°. The bearing of C from B is 080°. Mark the position of C on the grid. **(3 marks)**

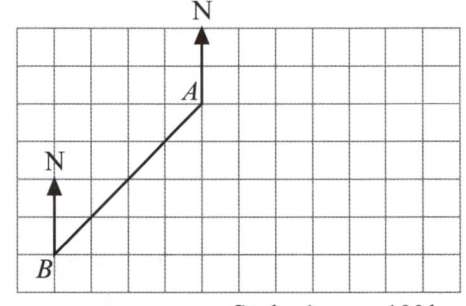

Scale: 1 cm = 100 km

102

Had a look ☐   Nearly there ☐   Nailed it! ☐

GEOMETRY & MEASURES

Circles

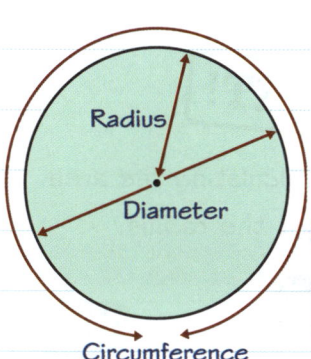

Make sure you know the definitions of **radius**, **diameter** and **circumference**.

Here are two different formulae for the circumference. You can use either.

Diameter = 2 × radius

1 Circumference = π × Diameter $C = \pi d$

2 Circumference = 2 × π × Radius $C = 2\pi r$

LEARN IT!

π

This symbol is the Greek letter 'pi'. It always stands for the same number.

π = 3.1415926...

Your calculator probably has a button for entering π into a calculation. You might need to press the SHIFT key first.

If your calculator leaves π in the answer, press the button to get your answer as a decimal.

You can also use 3.142 as the value of π in your exam.

Worked example   Target grade **3**

Work out the circumference of this circle. Give your answer to 2 decimal places.   **(2 marks)**

Circumference = $2\pi r$
= 2 × 3.142 × 6
= 37.704
= 37.70 cm (2 d.p.)

6 cm

Worked example   Target grade **4**

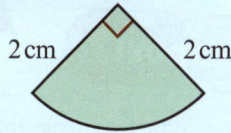

2 cm   2 cm

The diagram shows an earring made from a quarter of a circle.
Work out the perimeter of the earring.
Give your answer to 2 decimal places.   **(3 marks)**

Circumference of whole circle = $2\pi r$
= 2 × π × 2
= 12.5663...

Curved section of earring = 12.5663... ÷ 4
= 3.1415...

Total perimeter = 2 + 2 + 3.1415...
= 7.1415...
= 7.14 cm

Don't round your answers until the end of your calculation.

Now try this   Target grade **3**

This steering wheel has a circumference of 120 cm.
(a) Work out the diameter of the steering wheel.
Give your answer to 1 decimal place.   **(2 marks)**
(b) Work out the radius of the steering wheel.
Give your answer to 1 decimal place.   **(1 mark)**

Worked solution video

GEOMETRY & MEASURES

Had a look ☐   Nearly there ☐   Nailed it! ☐

Area of a circle

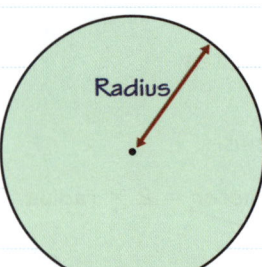

You need to know the formula for the area of a circle:

Area = π × Radius^2

$A = \pi \times r \times r = \pi r^2$

LEARN IT!

You always need to use the **radius** when you are calculating the area.
If you are given the diameter, divide it by 2 to get the radius.

Worked example *Target grade 3*

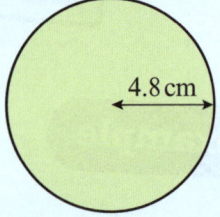

Work out the area of this circle.   **(2 marks)**

Area = πr^2

= $\pi \times 4.8^2$

= $\pi \times 23.04$

= 72.3822...

= 72.4 cm^2 (1 d.p.)

Worked example *Target grade 4*

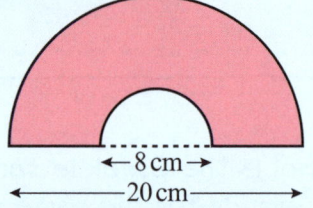

The diagram shows two semicircles with diameters 8 cm and 20 cm.
Work out the shaded area.   **(4 marks)**

Radius of large semicircle = 20 ÷ 2
 = 10

Area of large semicircle = $\frac{1}{2} \times \pi \times 10^2$
 = 157.0796...

Radius of small semicircle = 8 ÷ 2 = 4

Area of small semicircle = $\frac{1}{2} \times \pi \times 4^2$
 = 25.1327...

Shaded area = 157.0796... − 25.1327...
 = 131.946...
 = 132 cm^2 (3 s.f.)

Examiners' report

When you have to keep track of a lot of working, **write down** what each calculation represents. It will be really clear to you **and** the examiner what you have worked out!

> Real students have struggled with questions like this in recent exams – **be prepared!**

Examiners' report

The formula for the area of a circle uses the **radius**. If you are given the **diameter** you will need to divide by 2 first.

> Real students have struggled with questions like this in recent exams – **be prepared!**

Now try this *Target grade 4*

This circle and this square are equal in area.
Work out the length of the side of the square.
Give your answer correct to 1 decimal place.   **(4 marks)**

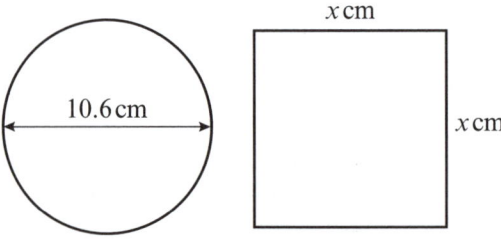

> Work out the area of the circle. This is the same as the area of the square. Work out the square root to find x.

Had a look ☐   Nearly there ☐   Nailed it! ☐

GEOMETRY & MEASURES

Tricky Topic

Sectors of circles

Each pair of radii divides a circle into two sectors, a **major sector** and a **minor sector**.

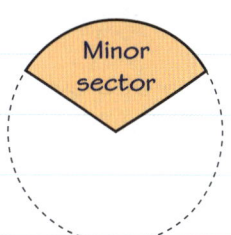
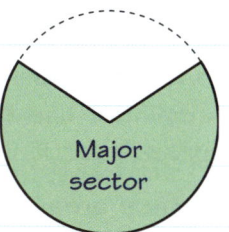

You can find the area of a sector by working out what fraction it is of the whole circle.

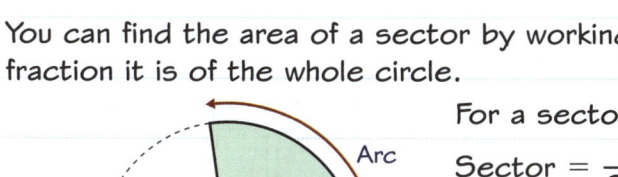

For a sector with angle x of a circle with radius r:

Sector = $\frac{x}{360°}$ of the whole circle so

Area of sector = $\frac{x}{360°} \times \pi r^2$

Arc length = $\frac{x}{360°} \times 2\pi r$

LEARN IT!

You can give answers in terms of π. There is more about this on the next page.

Worked example *Target grade 5*

The diagram shows a minor sector of a circle of radius 13 cm.

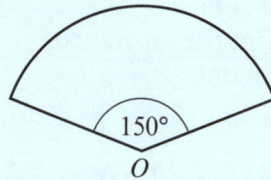

Work out the perimeter of the sector.   **(4 marks)**

Arc length = $\frac{x}{360°} \times 2\pi r$

= $\frac{150°}{360°} \times 2 \times \pi \times 13$

= 34.03392...

Perimeter = Arc length + Radius + Radius

= 34.03392... + 13 + 13

= 60 cm (2 s.f.)

Don't round until your final answer. The radius is given correct to 2 significant figures so this is a good degree of accuracy.

Finding a missing angle

You can use the formulae for arc length or area to find a missing angle in a sector. Practise this method to help you tackle the hardest questions.

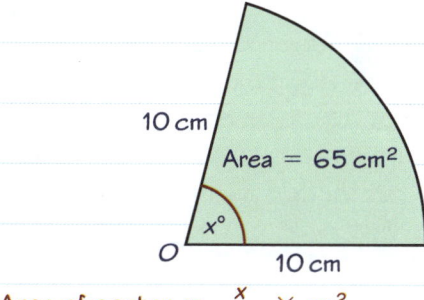

Area of sector = $\frac{x}{360} \times \pi r^2$

$65 = \frac{x}{360} \times 100\pi$

$x = \frac{65 \times 360}{100\pi}$

= 74.4845...

= 74.5° (to 3 s.f.)

Now try this *Target grade 5*

OAB is a sector of a circle, centre O.
Angle $AOB = 60°$.
$OA = OB = 12$ cm.
Work out the length of the arc AB.
Give your answer correct to
3 significant figures.   **(3 marks)**

You need to learn the formula for arc length.

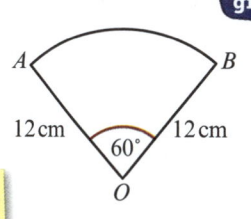

Cylinders

Surface area

To find the **surface area** of a **cylinder** you need to add up the areas of the faces. A cylinder has two flat circular faces and one curved face. When you flatten out the curved face it is shaped like a rectangle.

Surface area = 2 × Area of circle + Area of rectangle
         = 2 × πr^2 + $2\pi r$ × h
         = $2\pi r^2 + 2\pi rh$

LEARN IT!

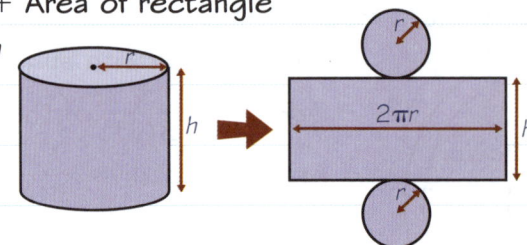

The length of the rectangular face is the same as the circumference of the circle.

Volume

For a cylinder with radius r and height h:

Volume of cylinder = Area of base × Height
                 = Area of circle × Height
                 = $\pi r^2 h$

LEARN IT!

Problem solved! You should always write down the formula **before** substituting. Be careful when deciding which quantities to use. You are given the **diameter** but the formula uses **radius** so you need to divide by 2.

Worked example

Target grade 4

This tin of soup is in the shape of a cylinder with height 11 cm. The diameter of the base is 7 cm.
$1 cm^3 = 1 ml$
Work out the capacity of the tin in ml. **(3 marks)**

Radius of base = 7 ÷ 2 = 3.5 cm

Volume = $\pi r^2 h$ = π × 3.5^2 × 11
       = 423.3296... cm^3

The capacity is 423 ml to the nearest whole number.

In terms of π

If a question asks for an **exact value** or an answer **in terms of π** then don't use the π button on your calculator. Write your answer as a whole number or fraction multiplied by π.

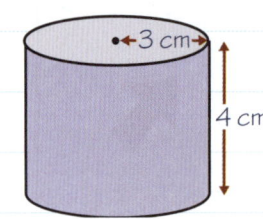

Volume of cylinder = $\pi r^2 h$
                  = π × 3^2 × 4

Exact answer
Volume = 36π cm^3

Rounded answer
Volume = 113 cm^3 (to 3 s.f.)

Now try this

Target grade 4

The diagram shows an oil drum in the shape of a cylinder of height 84 cm and diameter 58 cm.
It is one-quarter full of crude oil.
Calculate the volume of oil in the cylinder.
Give your answer in litres, correct to the nearest litre.
(4 marks)

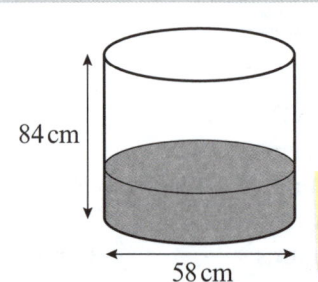

1 litre = 1000 cm^3

Had a look ☐ Nearly there ☐ Nailed it! ☐

GEOMETRY & MEASURES

Volumes of 3-D shapes

If you need to use these formulae in your exam they will be given to you with the question.

Cone

Volume of cone
$= \frac{1}{3} \times$ Area of $\times$ Vertical
         base   height
$= \frac{1}{3}\pi r^2 h$

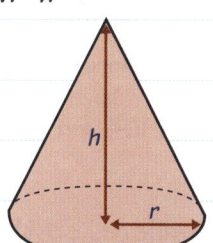

Sphere

Volume of sphere $= \frac{4}{3}\pi r^3$

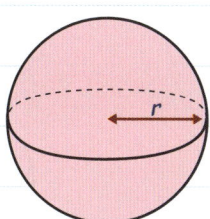

Pyramid

Volume of pyramid
$= \frac{1}{3} \times$ Area of $\times$ Vertical
         base   height
$= \frac{1}{3}Ah$

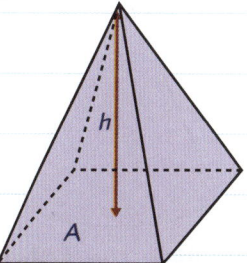

Worked example — Target grade 5

The diagram shows a cone A and a cylinder B. Show that the volume of B is 8 times the volume of A. **(4 marks)**

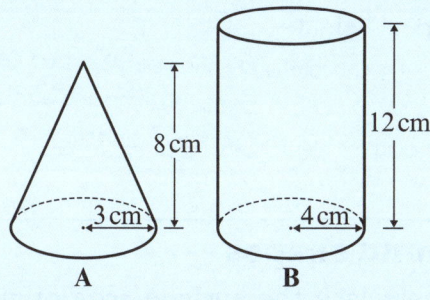

Volume of A $= \frac{1}{3}\pi r^2 h = \frac{1}{3}\pi \times 3^2 \times 8$
$\phantom{\text{Volume of A }}= 24\pi$

Volume of B $= \pi r^2 h = \pi \times 4^2 \times 12$
$\phantom{\text{Volume of B }}= 192\pi$

$8 \times 24\pi = 192\pi$ so the volume of B is 8 times the volume of A.

Examiners' report

You might have to **compare** two volumes or areas in your exam. These questions might involve:
- working out the ratio between two different areas or volumes
- finding an unknown quantity represented by a letter
- finding an expression for a length, area or volume in terms of an unknown.

In this question you need to know the ratio between the two volumes. Calculate them both, then write a short **conclusion**. Make sure you show the calculation in your conclusion: $8 \times 24\pi = 192\pi$
You can leave your working in terms of π to make it easier. There is more about this on page 106.

Real students have struggled with questions like this in recent exams – **be prepared!**

Now try this — Target grade 5

The radius of the base of a cone is 3 cm and its height is h cm.
The radius of a sphere is 3 cm.
The volume of the cone is equal to the volume of the sphere.
Find the value of h. You must show your working.

(3 marks)

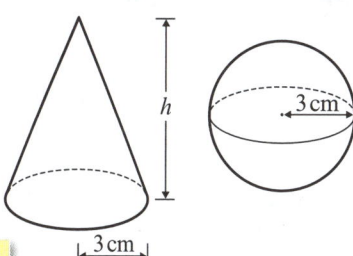

Write the expression for the volume of each shape; then set them equal to each other. Rearrange to make h the subject.

GEOMETRY & MEASURES

Tricky Topic

Had a look ☐ Nearly there ☐ Nailed it! ☐

Surface area

Cone

The formula for the **curved surface area** of a cone will be given if you need it for a question.

Curved surface area of cone = $\pi r l$

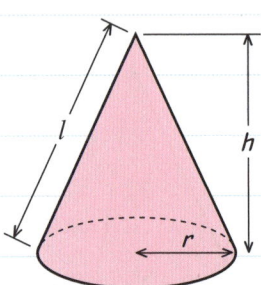

Be careful! This formula uses the slant height, l, of the cone.

To calculate the **total** surface area of the cone you need to add the area of the base. Surface area of cone = $\pi r^2 + \pi r l$

Sphere

The formula for the surface area of a sphere will be given if you need to use it.

Surface area of sphere = $4\pi r^2$

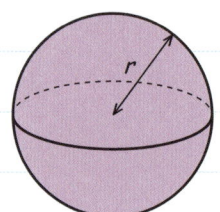

For a reminder about areas of circles and surface areas of cylinders have a look at pages 104 and 106.

A hemisphere is half a sphere, so the area of the curved surface is $\frac{1}{2} \times 4\pi r^2$

Worked example *Target grade 5*

The diagram shows a cone with vertical height 12 cm and base radius 5 cm. Work out the curved surface area of the cone. **(4 marks)**

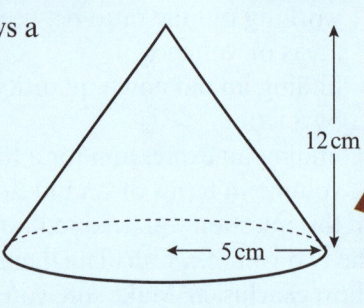

$l^2 = 12^2 + 5^2 = 169$
$l = 13$ cm
Curved surface area
$= \pi r l$
$= \pi \times 13 \times 5$
$= 65\pi \text{ cm}^2$

Problem solved! To work out the curved surface area you need to know the radius and the slant height. You are given the radius and the **vertical height**.

To calculate the slant height you need to use Pythagoras' theorem. Sketch the right-angled triangle containing the missing length.

Compound shapes

You can calculate the surface area of more complicated shapes by adding together the surface area of each part.

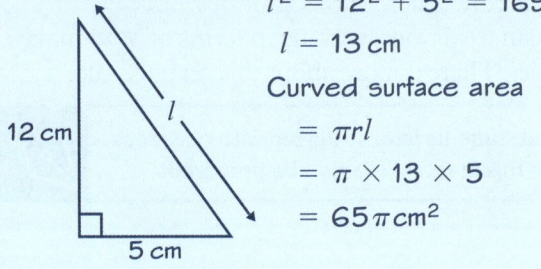

Surface area = $\pi(4)^2 + 2\pi(4)(6) + \frac{1}{2}[4\pi(4)^2]$
$= 96\pi \text{ cm}^2$

Now try this *Target grade 5*

1. The diagram shows an object made from two cones, placed one on top of the other.
   The top cone has a slant height of 6 cm and the bottom cone has a slant height of 7 cm.
   Both cones have a radius of 4 cm.
   Work out the total surface area of the object.
   Give your answer in terms of π. **(4 marks)**

   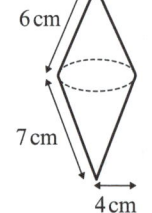

2. A solid object is formed by joining a hemisphere to a cylinder.
   Both the hemisphere and the cylinder have a radius of 2.1 cm. The cylinder has a height of 5.6 cm. Work out the total surface area of the object. Give your answer to 3 significant figures. **(4 marks)**

   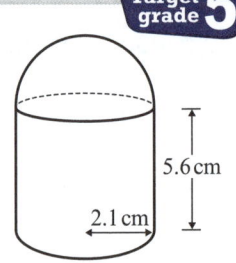

Had a look ☐   Nearly there ☐   Nailed it! ☐

GEOMETRY & MEASURES

Similarity and congruence

If one shape is an enlargement of another, the shapes are **similar**.

These triangles are similar.

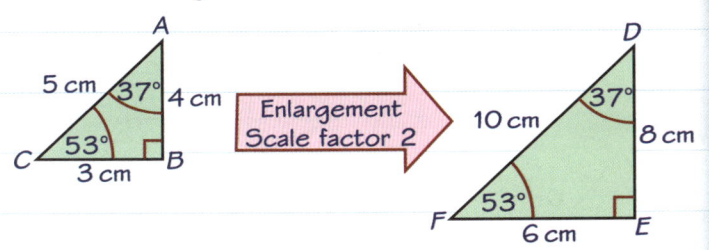

The angles in similar shapes are the same.

Congruent shapes are exactly the same shape and size. They have the same area and the same perimeter.

Rotations and reflections give congruent shapes. Enlargements are a different size so they are not congruent shapes.

Worked example — Target grade 1

Here is a large triangle made from 16 equilateral triangles.

Everything in blue is part of the answer.

Shade a shape which is **similar** to the shaded shape but **not** congruent. **(2 marks)**

Congruent shapes are exactly the same shape and size. Similar shapes can be enlargements. You can shade in a rhombus which is an enlargement of the grey shape with a scale factor of 2.

Worked example — Target grade 2

These two cubes are similar.

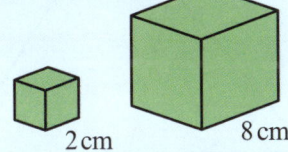

How many times will the 2 cm cube fit inside the 8 cm cube? **(2 marks)**

$4 \times 4 \times 4 = 64$

Problem solved! Imagine placing copies of the small cube inside the larger cube. A sketch might help.

You could fit $4 \times 4 = 16$ across the bottom.

You can fit 4 layers like this, so in total you can fit $4 \times 4 \times 4$ copies of the small cube inside the larger cube.

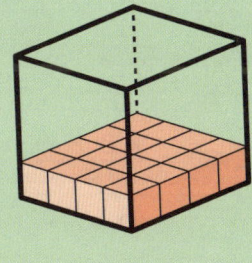

Now try this — Target grade 1

Here are some shapes on a grid.
Write down three pairs of similar shapes. **(3 marks)**

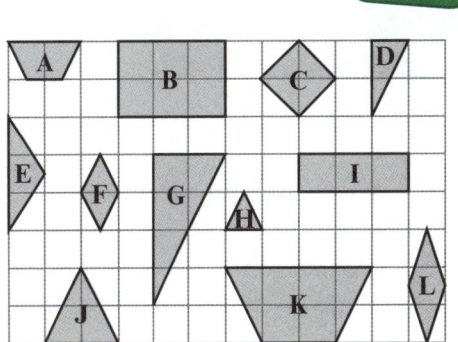

If one shape is an enlargement of another, the shapes are similar. Similar shapes are not necessarily congruent.

GEOMETRY & MEASURES Had a look ☐   Nearly there ☐   Nailed it! ☐

Similar shapes

You need to recognise similar triangles and find missing lengths and angles in similar shapes.
Similar triangles satisfy one of these three conditions:

1 All three pairs of angles are equal.
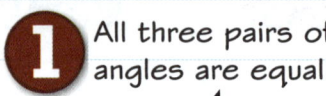

2 All three pairs of sides are in the same ratio.
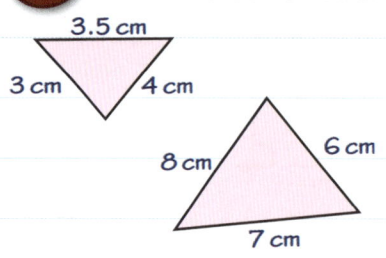

3 Two sides are in the same ratio and the included angle is equal.
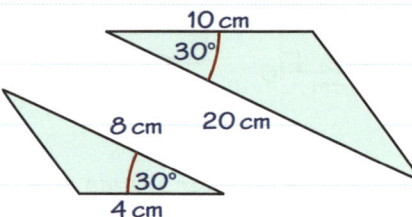

Worked example *Target grade 5*

XYZ and ABC are similar triangles.

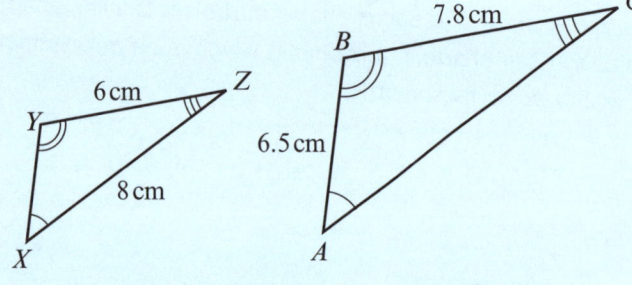

(a) Work out the length of AC. **(2 marks)**
(b) Work out the length of XY. **(2 marks)**

$\dfrac{AC}{XZ} = \dfrac{BC}{YZ}$

$\dfrac{AC}{8} = \dfrac{7.8}{6}$

$AC = \dfrac{7.8 \times 8}{6}$

$= 10.4\ cm$

$\dfrac{XY}{AB} = \dfrac{YZ}{BC}$

$\dfrac{XY}{6.5} = \dfrac{6}{7.8}$

$XY = \dfrac{6 \times 6.5}{7.8}$

$= 5\ cm$

> Start with the unknown length on top of a fraction. Make sure you write your ratios in the correct order.

Similar shapes checklist

Use these facts to solve similar shapes problems:

✓ Corresponding angles are equal.
✓ Corresponding sides are in the same ratio.

Spotting similar triangles

Here are some similar triangles:

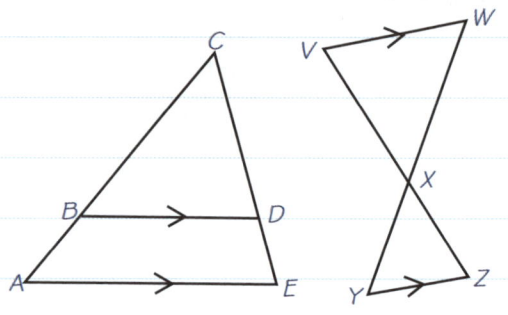

Triangle ACE is similar to triangle BCD.

Triangle VWX is similar to triangle ZYX.

Now try this *Target grade 5*

Triangles ABC and PQR are similar.
Angle ACB = angle PRQ.
(a) Work out the size of angle PRQ. **(2 marks)**
(b) Work out the length of PQ. **(2 marks)**

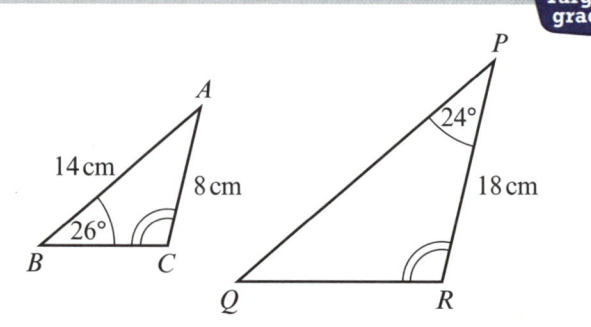

110

Had a look ☐ Nearly there ☐ Nailed it! ☐

GEOMETRY & MEASURES

Tricky Topic

Congruent triangles

Triangles are **congruent** if **any one** of these four conditions is true.

1 **SSS** (three sides are equal)

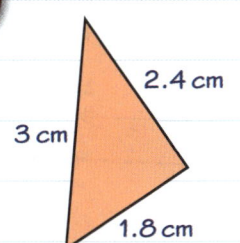

2 **AAS** (two angles and a corresponding side are equal)

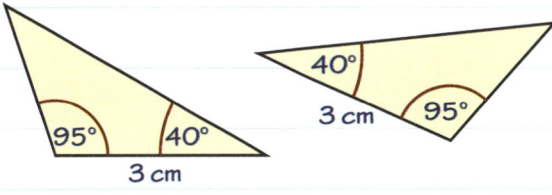

3 **SAS** (two sides and the included angle are equal)

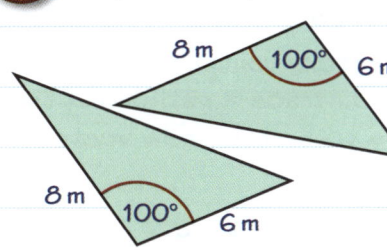

The angle must be **between** the two sides for SAS.

4 **RHS** (right angle, hypotenuse and a side are equal)

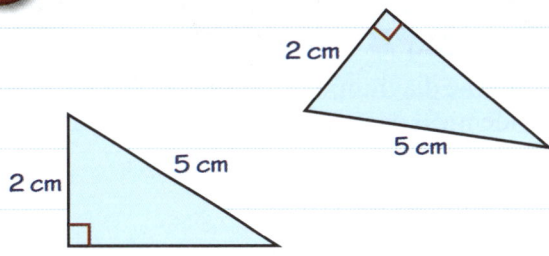

Worked example

Target grade 5

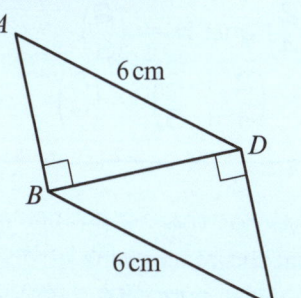

Triangle ABD and triangle BDC are congruent. Explain which information in the diagram supports this statement. **(3 marks)**

ABD and BDC are both right-angled triangles. The hypotenuse of both triangles is 6 cm. Side BD is common to both triangles so is equal. Therefore the triangles satisfy the RHS condition and are congruent.

Problem solved! To explain why the two triangles are congruent, you need to show that **one** of the conditions above is satisfied. Once you have decided which one, explain which lengths or angles are the same and why. Then write down **which condition** is satisfied. It's OK to use the abbreviations above.

Common sides

If two triangles have a side in **common** then those two sides are equal.

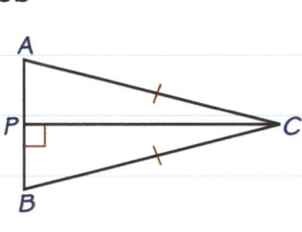

Now try this

Target grade 5

In the quadrilateral $ABCD$, $BC = CD$ and the angles at B and D are right angles.

Show that triangle ABC is congruent to triangle ADC. **(3 marks)**

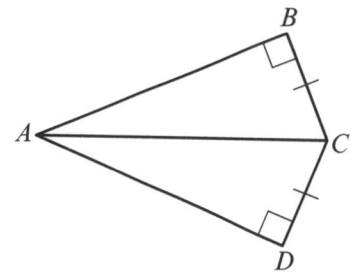

111

GEOMETRY & MEASURES

Had a look ☐   Nearly there ☐   Nailed it! ☐

Tricky Topic

Vectors

A vector has a **magnitude** (or size) and a **direction**.

This vector can be written as
a, $\vec{AB}$ or $\begin{pmatrix} 2 \\ 5 \end{pmatrix}$

You can multiply a vector by a number. The new vector has a different length but the same direction.

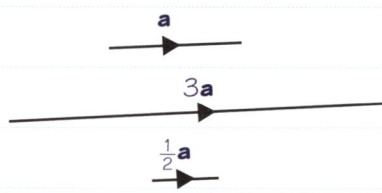

If **b** is a vector then −**b** is a vector with the same length but opposite direction.

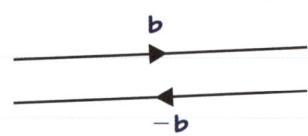

Worked example   *Target grade 5*

In the diagram, *OADB* and *ACED* are two identical parallelograms.

a is the vector $\vec{OA}$
b is the vector $\vec{OB}$

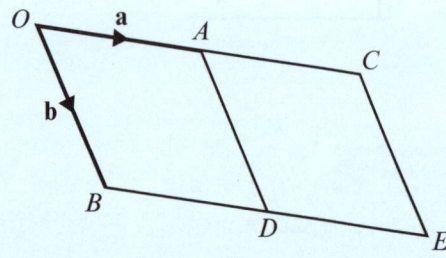

Find the following vectors in terms of **a** and **b**.

(a) $\vec{OD}$ **(1 mark)**

$\vec{OD} = \vec{OA} + \vec{AD}$
$= a + b$

(b) $\vec{EB}$ **(1 mark)**

$\vec{EB} = \vec{ED} + \vec{DB}$
$= -a + -a = -2a$

(c) $\vec{BC}$ **(1 mark)**

$\vec{BC} = \vec{BD} + \vec{DE} + \vec{EC}$
$= a + a - b = 2a - b$

Adding vectors

You can add vectors using the **triangle law**. You trace a path along the added vectors to find the new vector.

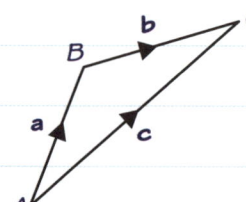

a + **b** = **c**

c is the resultant vector of **a** and **b**.

If $a = \begin{pmatrix} 2 \\ 4 \end{pmatrix}$ and $b = \begin{pmatrix} 6 \\ 3 \end{pmatrix}$

then $c = \begin{pmatrix} 2 + 6 \\ 4 + 3 \end{pmatrix} = \begin{pmatrix} 8 \\ 7 \end{pmatrix}$

> For each vector, trace a path along the shape from the start point to the end point. If you go in the **opposite** direction to the vector then you need to **subtract**.
> $\vec{AD} = \vec{OB}$ because they are **parallel**.

> Simplify your vectors as much as possible.

> If you multiply a column vector by a number you have to multiply **both parts**.
> $2 \times \begin{pmatrix} p \\ q \end{pmatrix} = \begin{pmatrix} 2p \\ 2q \end{pmatrix}$
> To **add** column vectors you add the top numbers and add the bottom numbers.
> $\begin{pmatrix} c \\ d \end{pmatrix} + \begin{pmatrix} e \\ f \end{pmatrix} = \begin{pmatrix} c + e \\ d + f \end{pmatrix}$

Now try this   *Target grade 5*

The vectors **a** and **b** are defined as

$a = \begin{pmatrix} 3 \\ 5 \end{pmatrix}$ $b = \begin{pmatrix} 2 \\ -9 \end{pmatrix}$

Write the following as column vectors.

(a) 2**a**   **(1 mark)**
(b) **a** + **b**   **(1 mark)**
(c) **b** − 3**a**   **(3 marks)**

Had a look ☐   Nearly there ☐   Nailed it! ☐

GEOMETRY & MEASURES

Problem-solving practice 1

About half of the questions in your Foundation GCSE exam will require you to **problem-solve**, **reason**, **interpret** or **communicate** mathematically. If you come across a tricky or unfamiliar question in your exam you can try some of these strategies.

- ✓ Sketch a diagram to see what is going on.
- ✓ Try the problem with smaller or easier numbers.
- ✓ Plan your strategy before you start.
- ✓ Write down any formulae you might be able to use.
- ✓ Use x or n to represent an unknown value.

AO2

AO3

Now try this

1 This diagram shows a quadrilateral.

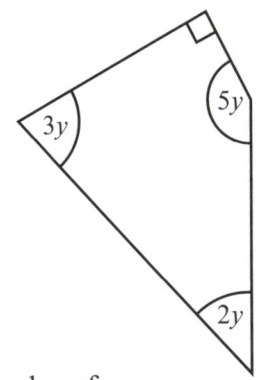

Work out the value of y.   **(3 marks)**

Angles 2 page 74
Using algebra page 51

Target grade **3**

Start by writing an expression for the sum of the angles in the quadrilateral. Use angle facts to write down what this sum is equal to. This gives you an equation you can solve to find the value of y.

TOP TIP

If you know what unknown angles should add up to, then you can write an equation and solve it to find the value of the unknown.

2 A buoy is 6 km from a ship on a bearing of 290°.
A lighthouse is 8 km east of the ship. Work out the distance between the buoy and the lighthouse.   **(3 marks)**

Bearings page 102
Scale drawings and maps page 98

Target grade **3**

You'll need to draw a scale diagram to solve this problem. A good scale to use would be 1 cm = 2 km. Always use a ruler and a sharp pencil to draw any lines and **don't** rub out any construction lines or working.

TOP TIP

Bearings less than 180° are to the **right** of north. Bearings between 180° and 360° are to the **left**. Bearings always have three figures.

Problem-solving practice 2

Now try this

3 The diagram shows a semicircle. It has two identical semicircles drawn inside it.

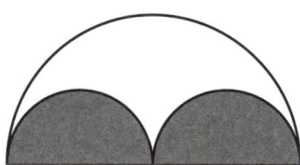

Show that the shaded area is exactly half of the area of the large semicircle. **(3 marks)**

Area of a circle page 104 *Target grade 3*

The easiest way to tackle this problem is to choose a value for the diameter of each semicircle. The two small semicircles are identical, so if they both have diameter 2 cm, the large semicircle will have diameter 4 cm. Work out the area of each semicircle using these values, then write a conclusion. Make sure all your working is clear and neat.

TOP TIP

If you aren't given any lengths on a diagram you might be able to:
- use x to represent one of the lengths and write the other lengths in terms of x
- choose some easy numbers yourself for the lengths.

---

4 A ladder is 6 m long.
The ladder is placed on horizontal ground, resting against a vertical wall.
The instructions for using the ladder say that the bottom of the ladder must not be closer than 1.5 m to the bottom of the wall.
How far up the wall can the ladder reach if the instructions are followed? **(3 marks)**

Pythagoras' theorem page 90 *Target grade 4*

You should definitely draw a sketch to show the information in the question.

TOP TIP

Be careful when you are working out the length of a **short** side using Pythagoras' theorem.
Remember: $\text{short}^2 + \text{short}^2 = \text{long}^2$
$\text{short}^2 = \text{long}^2 - \text{short}^2$

5

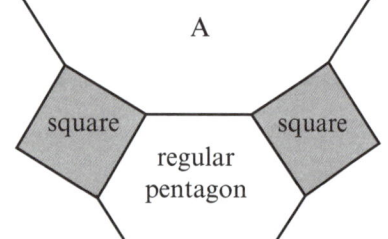

The diagram shows two squares, part of a regular pentagon and part of a regular n-sided polygon, A.
Calculate the value of n. Show your working clearly. **(5 marks)**

Angles in polygons page 76 *Target grade 4*

Follow these steps:
1. Work out the interior angles of a regular pentagon and a square.
2. Use the fact that the angles around a point add up to 360° to find the interior angle of A.
3. Subtract this from 180° to find the exterior angle of A.
4. Divide 360° by this to work out n.

TOP TIP

If there are a lot of steps in a question it's a good idea to **plan** your answer before you start.

Had a look ☐ Nearly there ☐ Nailed it! ☐

PROBABILITY & STATISTICS

Two-way tables

You can answer questions about two-way tables by adding or subtracting.

	Year 7	Year 8	Year 9	Total
Vegetarian	14	22	25	61
Not vegetarian	72	63	54	189
Total	86	85	79	250

There were 61 vegetarians in total.

In total 250 students were surveyed.

There were 86 Year 7 students surveyed.

There were 63 non-vegetarians in Year 8.

Worked example

Target grade 3

Anton surveyed 120 people about how they voted at the last general election. He recorded the results in a two-way table:

	Labour	Conservative	Other	Total
Female	21	13	13	47
Male	32	27	14	73
Total	53	40	27	120

Complete the two-way table. **(4 marks)**

Labour column: 53 − 21 = 32
Female row: 47 − 21 − 13 = 13
Conservative column: 13 + 27 = 40
Total row: 120 − 53 − 40 = 27
Other column: 27 − 13 = 14
Male row: 32 + 27 + 14 = 73
Check: 47 + 73 = 120
53 + 40 + 27 = 120 ✓

Everything in blue is part of the answer.

Golden rules

The numbers in each column add up to the total for that column.

The numbers in each row add up to the total for that row.

Other
13
+ 14
= 27

Female	21	+ 13	+ 13	= 47

To complete a two-way table:
- Write the total in the bottom right-hand cell.
- Look for rows and columns with only one missing number.
- Use addition and subtraction to find any missing values.
- Fill in the missing values as you go along.

Check it!
Add up the row totals and the column totals. They should be the same.

Now try this

Target grade 3

A photographic shop offers prints in three different sizes and on three different types of paper.

This two-way table shows information about the choices made by customers on one day.

Complete the two-way table. **(4 marks)**

Worked solution video

	Gloss	Matt	Lustre	Total
Small	20	35	12	
Medium	63		29	
Large		24		
Total	105		59	325

Look for rows or columns with only one missing number.

115

Pictograms

A **pictogram** can be used to represent data from a tally chart or frequency table. This pictogram shows the results of a survey about how people watch television. There is one row for each option.

Key: 📺 represents 2 people

*A pictogram must have a **key**. This tells you how many items are represented by each picture.*

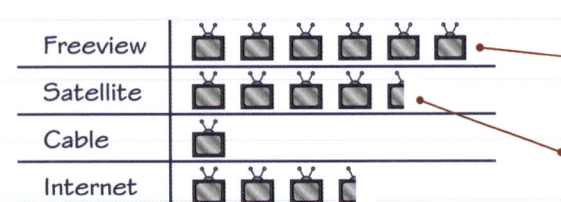

📺 represents 2 people so 12 people said they watched television using Freeview.

Each television represents 2 people, so half a television represents 1 person. This row represents 9 people.

To work out the total number of people in the survey, add together the totals of each row: $12 + 9 + 2 + 7 = 30$.

Worked example (Target grade 1)

The pictogram shows the numbers of films Adam watched in June, July and August.

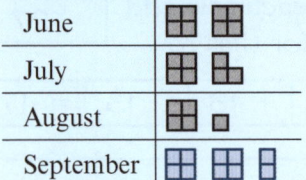

Key: ⊞ represents 4 films

(a) Write down the number of films Adam watched in June. **(1 mark)**

8

(b) Write down the number of films Adam watched in August. **(1 mark)**

5

In September Adam watched 10 films.
(c) Use this information to complete the pictogram. **(1 mark)**

Everything in blue is part of the answer.

Use the key to work out what each picture represents.

⊞ = 4 films   ⌐⊞ = 3 films
▯▯ = 2 films   ▫ = 1 film

There is a block of 4 squares and a block of 1 square in August. This represents $4 + 1 = 5$ films.

To represent 10 films you need two blocks of 4 squares and one block of 2 squares. Draw the squares neatly on the pictogram.

Tallies

A **tally** is a bit like a pictogram:
| represents 1 ||||| represents 5
So to represent 12 you would draw:
|||| |||| ||

Now try this (Target grade 2)

A magazine uses a pictogram to show the average house price in five different English counties.
The key for the pictogram is missing.
The average house price in Herefordshire is £225 000.
The average house price in Lincolnshire is £125 000.

(a) Complete the pictogram to show this information. Include a key for the pictogram. **(2 marks)**

(b) Work out the average house price in Essex. **(1 mark)**

Worked solution video

Key

County	Houses
Lancashire	🏠🏠🏠
Herefordshire	🏠🏠🏠🏠⌐
Essex	🏠🏠🏠🏠🏠
Surrey	🏠🏠🏠🏠🏠🏠🏠⌐
Lincolnshire	

Had a look ☐   Nearly there ☐   Nailed it! ☐

PROBABILITY & STATISTICS

Bar charts

You can use a **bar chart** to represent data given in a tally chart or frequency table.
This **dual bar chart** shows the numbers of pairs of jeans owned by the students in a class.

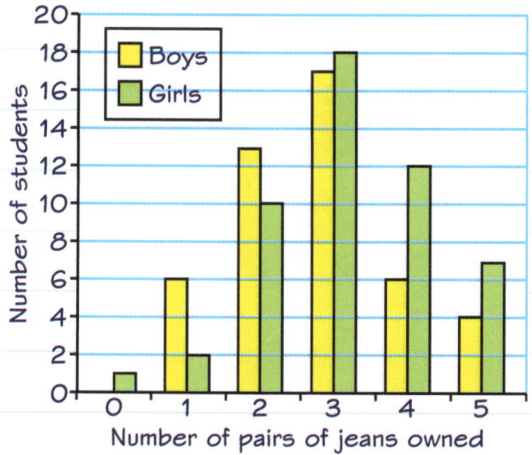

Bar chart features
✓ Bars are the same width.
✓ There is a gap between the bars.
✓ Both axes have labels.
✓ Bars can be drawn horizontally or vertically.
✓ The height (or length) of each bar represents the frequency.
✓ In a dual bar chart two (or more) bars are drawn side by side. They can be used to compare data.

Worked example

 Target grade 1

Kaitlyn carried out a survey of the colours of cars which passed the school gate in 10 minutes. Here are her results.

Colour	Tally
Red	|||| ||||
Black	|||| ||
Silver	|||| ||||
Blue	|||

Use the grid to draw a <u>suitable chart or diagram</u> to represent Kaitlyn's results.   **(4 marks)**

The question says 'suitable chart or diagram'. You could also represent this data using a pictogram or pie chart.

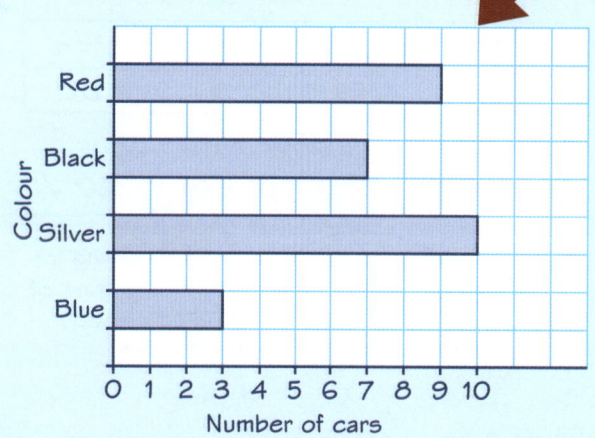

Now try this

Worked solution video

The bar chart shows the number of job applications Susan sent during one week.
On Tuesday Susan sent 12 job applications.
How many job applications did Susan send altogether during this week?
You must show your working.   **(3 marks)**

 Target grade 1

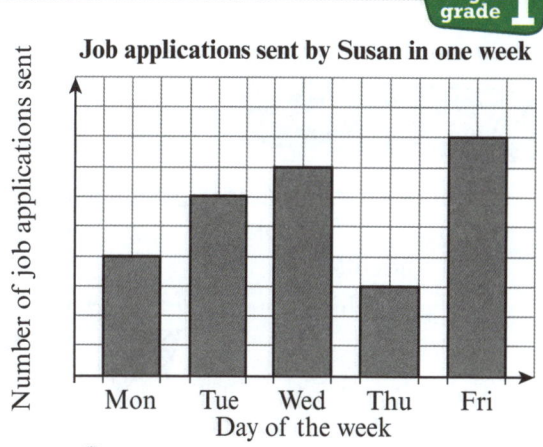

Start by using the information in the question to work out the scale on the vertical axis.

117

PROBABILITY & STATISTICS

Had a look ☐ Nearly there ☐ Nailed it! ☐

Pie charts

In your exam you might have to interpret information given in a **pie chart**, or draw a pie chart from a frequency table. This pie chart shows the favourite sports of 120 students.

There are 360° in a circle and 120 students.
360 ÷ 120 = 3
This means that each student is represented by an angle of 3°.

66 ÷ 3 = 22
So 66° represents 22 students.

Cricket 30°
Tennis 66°
Football 180°
Rugby 84°

Half the students chose football. This **sector** of the pie chart represents 60 students.

28 students said their favourite sport was rugby.

Worked example (Target grade 2)

A farm has 40 fruit trees.
The table shows the number of each type of tree.
Draw a pie chart to represent this information.

Type of fruit tree	Number of trees	Angle
apple	12	12 × 9° = 108°
plum	5	5 × 9° = 45°
pear	14	14 × 9° = 126°
peach	9	9 × 9° = 81°

Angle for 1 tree = 360° ÷ 40 = 9°

Check: 108° + 45° + 126° + 81° = 360° ✓

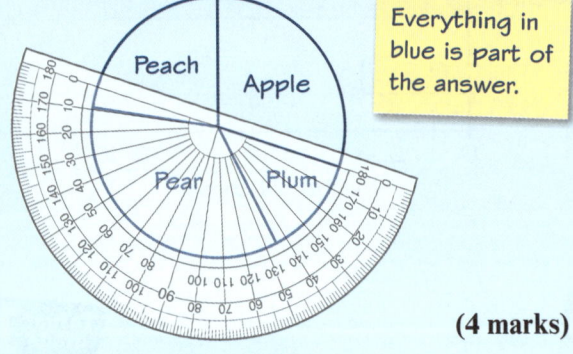

Everything in blue is part of the answer.

(4 marks)

You can revise measuring and drawing angles using a protractor on page 95.

You need a sharp pencil, compasses and a protractor to draw a pie chart.

1. Add an 'Angle' column to the frequency table.
2. There are 360° in a full circle. There are 40 trees. So divide 360° by 40 to find the angle that represents 1 tree.
3. Multiply the angle that represents 1 tree by the number of each tree type to find the angle for each type.
4. **Check** that your angles add up to 360°.
5. Draw a circle using compasses. Draw a vertical line from the centre to the edge of the circle. Use a protractor to measure and draw the first angle (108°) from this line. Draw each angle carefully in order.
6. Label each sector of your pie chart with the type of fruit tree.

Now try this (Target grade 2)

A school canteen offered four lunch options on one day. This table shows the choices made by 90 students.

Draw a pie chart to represent this data.

(4 marks)

Start by drawing a circle with a radius of at least 3 cm.

Lunch choice	Number of students
salad	21
stew	9
pizza	45
curry	15

Had a look ☐   Nearly there ☐   Nailed it! ☐

PROBABILITY & STATISTICS

Scatter graphs

The points on a scatter graph aren't always scattered. If the points are almost on a straight line then the scatter graph shows **correlation**. The better the straight line, the stronger the correlation.

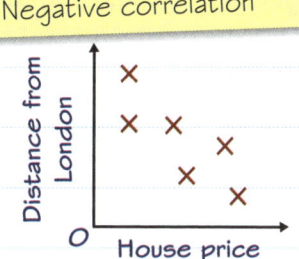
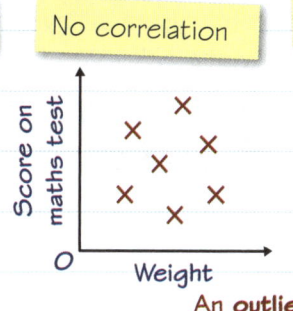
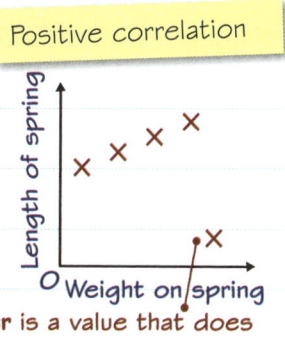

An **outlier** is a value that does not fit the pattern of the data.

Worked example

This scatter graph shows the engine capacity of some cars and the distance they will travel on one gallon of petrol.

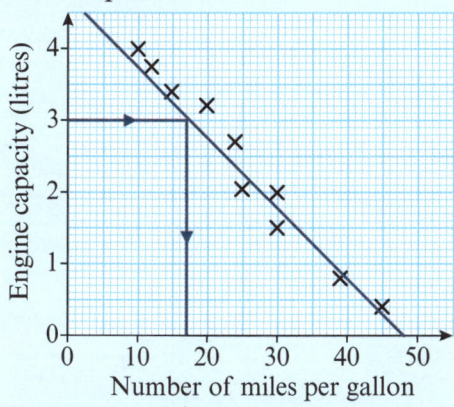

(a) Estimate how far a car with a 3-litre engine will travel on 1 gallon of petrol. **(2 marks)**

17 miles

(b) Comment on the reliability of your estimate. **(1 mark)**

3 litres is within the range of the data values (interpolation) and the correlation is strong, so the estimate is reliable.

Cause and effect

Watch out! Correlation doesn't always mean that two variables are related.

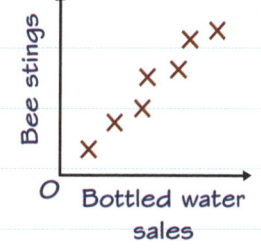

Bottled water doesn't cause bee stings but, when the weather is hotter, bottled water sales and bee stings both increase.

Use a **line of best fit** to estimate. Draw a straight line as close to as many of the points on the graph as possible.

Estimating and predicting

✓ If you are predicting a value that is **within** the range of the data your prediction will be **more accurate**.
This is called **interpolation**.

✗ If you are predicting a value that falls **outside** the range of the data your prediction will be **less accurate**.
This is called **extrapolation**.

Now try this

This scatter graph shows the daily hours of sunshine and the daily maximum temperature at 10 seaside resorts in England one day last summer.

(a) Another resort had 10 hours of sunshine each day. Use a line of best fit to predict the maximum temperature at this resort. **(2 marks)**

(b) Comment on the reliability of your estimate. **(1 mark)**

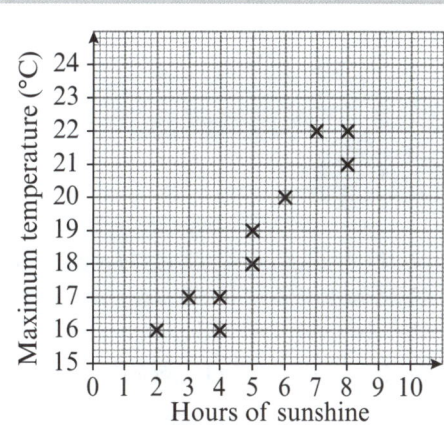

PROBABILITY & STATISTICS

Had a look ☐   Nearly there ☐   Nailed it! ☐

Averages and range

There are three different types of average: the mean, median and mode.
The range of a set of data tells you how spread out the data is.

Worked example — Target grade 1

The **mode** is the value which occurs most often.

Here are six numbers:
4 5 9 7 4 4

(a) Write down the mode. **(1 mark)**
The mode is 4.

(b) Work out the mean. **(2 marks)**
$4 + 5 + 9 + 7 + 4 + 4 = 33$
$33 \div 6 = 5.5$
The mean is 5.5

The **median** is the middle value. First write the values in order from smallest to largest. If there are two middle values, the median is halfway between them.

(c) Work out the median. **(2 marks)**
4 4 ④ ⑤ 7 9
The median is 4.5

(d) Work out the range. **(1 mark)**
$9 - 4 = 5$
The range is 5

Examiners' report

Make sure you know the difference between the **mean**, the **median** and the **mode**.

Real students have struggled with questions like this in recent exams – **be prepared!**

To find the **mean** you add together all the numbers and then divide by how many numbers there are. Don't round your answer.

Range = Largest value − Smallest value

Worked example — Target grade 3

Kayla has three numbered cards. The numbers are hidden.

The mode of the three numbers is 5.
The mean of the three numbers is 4.
Work out the three numbers on the cards. **(3 marks)**

Mode = 5
At least two of the cards are 5s.
Mean = 4
Sum of three cards = $4 \times 3 = 12$
$5 + 5 + ? = 12$
The other card is a 2.
The cards are 5, 5 and 2.

Problem solved! The mode is the most common value. There are three cards so at least two must have 5 written on them.

To find the other value you can use this formula:

Sum of values = Mean × Number of values

Check it!
Calculate the mean of your three values:
$5 + 5 + 2 = 12$
$12 \div 3 = 4$ ✓

Now try this — Target grade 1

Mrs Miller gives 11 students in her class a mental arithmetic test.
The test is marked out of 15. These are their marks:
   12 13 8 14 11 8 13 14 15 14 10

(a) Write down the mode. **(1 mark)**   (b) Work out the range. **(1 mark)**
(c) Work out the median. **(2 marks)**   (d) Work out the mean. **(2 marks)**

Had a look ☐ Nearly there ☐ Nailed it! ☐

PROBABILITY & STATISTICS

Averages from tables 1

You need to be really careful when you are calculating averages from a frequency table. Remember that the **frequency** is not a data value – it tells you the **number** of pieces of data with a given value.

Worked example

Target grade 1

Leah asked 40 people how many pets they owned. The table shows her results.

Number of pets x	Frequency f	Frequency × number of pets $f \times x$
0	12	$12 \times 0 = 0$
1	18	$18 \times 1 = 18$
2	7	$7 \times 2 = 14$
3	2	$2 \times 3 = 6$
4	1	$1 \times 4 = 4$

(a) Write down the mode. **(1 mark)**
The mode is 1 pet.

(b) Write down the range. **(1 mark)**
The range is $4 - 0 = 4$ pets.

(c) Work out the median. **(1 mark)**
The median is 1 pet.

(d) Work out the mean. **(3 marks)**
Total of $f \times x$ column $= 0 + 18 + 14 + 6 + 4$
$= 42$
Total frequency $= 12 + 18 + 7 + 2 + 1 = 40$
$42 \div 40 = 1.05$
The mean is 1.05 pets.

Everything in blue is part of the answer.

(a) The **mode** is the value with the highest frequency. The highest frequency is 18.

(b) The **range** is the difference between the highest and lowest values. So $4 - 0 = 4$ pets.

Examiners' report

(c) To find the **median** from a frequency table you need to look at the **frequencies**. There are 40 values so the median is between the 20th and 21st values. The first 12 values are all 0, and the next 18 are all 1. This means the 20th and 21st values are both 1, so the median is 1 pet.

Real students have struggled with questions like this in recent exams – **be prepared!**

(d) To calculate the **mean** from a frequency table you need to add an extra column.

The total in the $f \times x$ column represents the total number of pets owned (42 pets).

Use this rule to work out the mean:

$$\text{Mean} = \frac{\text{Total of } (f \times x) \text{ column}}{\text{Total frequency}}$$

Do not round your answer.

Now try this

Target grade 1

The table shows the number of goals scored by 50 teams in matches taking place one weekend.

(a) Write down the modal number of goals scored. **(1 mark)**
(b) Work out the median number of goals scored. **(1 mark)**
(c) Work out the mean number of goals scored. **(3 marks)**

Goals scored (x)	Frequency (f)	
0	7	
1	15	
2	14	
3	7	
4	4	
5	3	

Use the extra column for 'Frequency × Goals scored ($f \times x$)'.

121

PROBABILITY & STATISTICS   Had a look ☐   Nearly there ☐   Nailed it! ☐

Averages from tables 2

Sometimes data in a frequency table is grouped into **class intervals**. You don't know the exact data values, but you can calculate an **estimate** of the mean, and write down which class interval contains the median and which one has the highest frequency.

Worked example

Target grade 4

Maisie recorded the times, in minutes, taken by 150 students to travel to school. The table shows her results.

Time (t minutes)	Frequency f	Mid-point x	$f \times x$
$0 \leq t < 20$	65	10	$65 \times 10 = 650$
$20 \leq t < 40$	42	30	$42 \times 30 = 1260$
$40 \leq t < 60$	39	50	$39 \times 50 = 1950$
$60 \leq t < 80$	4	70	$4 \times 70 = 280$

Everything in blue is part of the answer.

(a) Write down the modal class interval. **(1 mark)**

$0 \leq t < 20$

(b) Write down the class interval which contains the median. **(1 mark)**

$20 \leq t < 40$

(c) Work out an estimate for the mean number of minutes that the students took to travel to school. **(4 marks)**

Sum of $f \times x$ column
$= 650 + 1260 + 1950 + 280 = 4140$

Total frequency
$= 65 + 42 + 39 + 4 = 150$

$4140 \div 150 = 27.6$

Estimated mean = 27.6 minutes.

1. Add extra columns to the table.
2. Use the first extra column for 'mid-point' and work out the mid-point of each class interval.
3. Label 'frequency' f and 'mid-point' x.
4. Use the final column for $f \times x$. Make sure you use the mid-point when calculating $f \times x$ for each row.
5. Use this rule to estimate the mean.

$$\text{Estimate of mean} = \frac{\text{Total of } (f \times x \text{ column})}{\text{Total frequency}}$$

(d) Explain why your answer to part (c) is an estimate. **(1 mark)**

Because you don't know the exact data values.

Now try this

Target grade 4

The table shows the marks obtained by 50 students in a maths test.

(a) Write down the modal class interval. **(1 mark)**
(b) Write down the class interval that contains the median. **(1 mark)**
(c) Work out an estimate of the mean mark for these students. **(4 marks)**

Marks (x)	Frequency (f)
$0 < x \leq 10$	10
$10 < x \leq 20$	11
$20 < x \leq 30$	8
$30 < x \leq 40$	6
$40 < x \leq 50$	10
$50 < x \leq 60$	5

Had a look ☐ Nearly there ☐ Nailed it! ☐

PROBABILITY & STATISTICS

Line graphs

A **time series** graph is a line graph that shows how a variable changes over a period of time. This time series graph shows median incomes in the UK from 1980 to 2010.

The graph should have a title. • — • UK median incomes

The **vertical axis** shows the variable which changes over time.

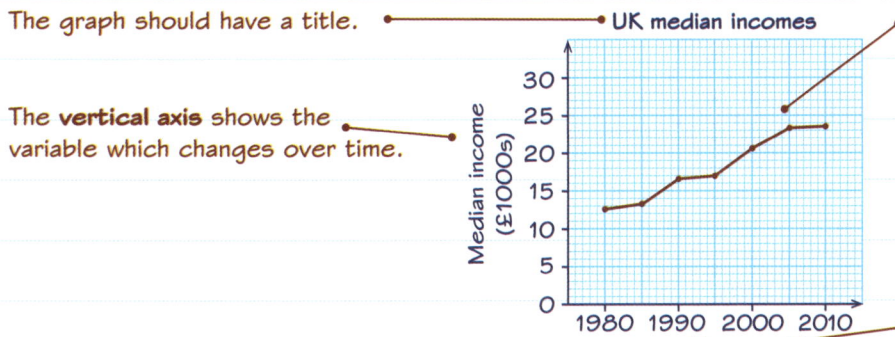

The graph shows an **upward trend**. This means that in general, as time passes, the median income **increases**.

The **horizontal axis** always shows units of time.

Vertical line graphs

A vertical line graph is sometimes called a **bar-line graph**. It works in the same way as a **bar chart** but the bars are straight, vertical lines.

You can revise bar charts on page 117.

Worked example

This vertical line graph shows the number of pairs of trainers owned by the students in a class.

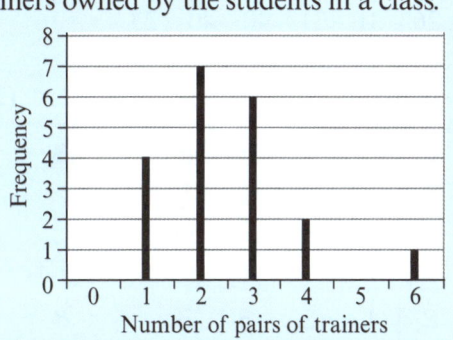

What was the mean number of pairs of trainers owned? **(2 marks)**

Total number of pairs of trainers owned
$0 \times 0 + 4 \times 1 + 7 \times 2 + 6 \times 3 + 2 \times 4 + 0 \times 5 + 1 \times 6 = 50$

Total number of students in the class
$0 + 4 + 7 + 6 + 2 + 0 + 1 = 20$
$50 \div 20 = 2.5$
The mean is 2.5 pairs of trainers.

 This is like finding the mean from this frequency table:

Number of pairs of trainers	0	1	2	3	4	5	6
Frequency	0	4	7	6	2	0	1

Work out the total number of pairs of trainers owned, then divide by the total number of students in the class.

You can revise how to find the mean from a frequency table on page 121.

Now try this

Target grade 4

The table on the right gives information about the percentage of students who passed a professional exam each year from 2007 to 2013.

(a) Draw a time series graph to represent this data. **(3 marks)**
(b) Describe the trend. **(1 mark)**

Year	2007	2008	2009	2010	2011	2012	2013
Percentage (%)	58	60	53	55	63	65	73

You can use a 'break' in your vertical axis.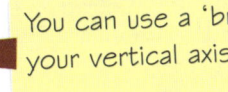

123

PROBABILITY & STATISTICS

Had a look ☐   Nearly there ☐   Nailed it! ☐

Stem-and-leaf diagrams

When data is given in a **stem-and-leaf diagram** it is arranged in order of size.
This stem-and-leaf diagram shows the costs, in £, of some DVDs.

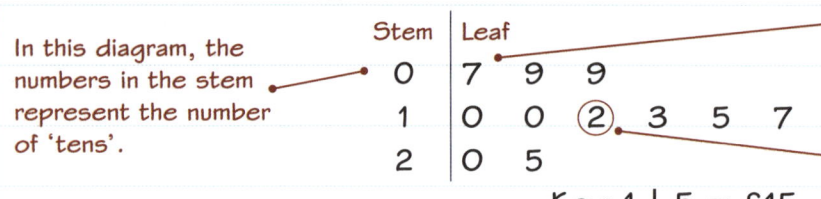

In this diagram, the numbers in the stem represent the number of 'tens'.

Stem | Leaf
0 | 7 9 9
1 | 0 0 ② 3 5 7
2 | 0 5

Key: 1 | 5 = £15

0 | 7 represents £7. This is the smallest data value.

There are 11 pieces of data, so the median is the 6th value. The median is £12.

There are 11 pieces of data in this stem-and-leaf diagram.
The range of prices is £25 − £7 = £18.

Worked example — Target grade 1

Alison records the time in minutes it takes her to drive to work each morning.

35 28 21 23 33 23 44 36
41 26 31 31 40 40 23

(a) Show the data in an ordered stem-and-leaf diagram. **(3 marks)**

2 | 8 1 3 3 6 3
3 | 5 3 6 1 1
4 | 4 1 0 0

Key: 2|1 represents 21 minutes

2 | 1 3 3 3 6 8
3 | 1 1 3 5 6
4 | 0 0 1 4

(b) Work out the median time. **(1 mark)**

31 minutes

(c) Write down the modal time. **(1 mark)**

23 minutes

(d) Alison says the range of times is 3 minutes because 4 − 1 = 3. Explain why Alison is wrong. **(1 mark)**

44 − 21 = 23
The range is actually 23 minutes.

To draw a stem-and-leaf diagram:

1. Choose sensible values to use as your stem.
2. Draw an ordered stem, then add the leaves in any order.
3. Cross each data value off the list as you enter it.
4. Redraw the diagram, putting the leaves in order.
5. Add a key.

Averages

Median — the data is already in order, so it's easy to find the middle value.
Watch out! — You need to give the whole data value, not just the leaf.

Mode — look for the most commonly occurring leaf in the **same** row.
Watch out! — The same leaf represents different values in different rows.

Range — subtract the lowest value from the highest value.
Watch out! — Use the actual data values, not just the value of the stems.

Now try this — Target grade 1

Seventeen people recorded the time it took them to solve a puzzle.
Here are their times in minutes:

24 12 9 16 24 8 26 19 24
28 18 10 23 15 11 9 12

(a) Complete an ordered stem-and-leaf diagram to represent this data. **(3 marks)**
(b) What is the modal time? **(1 mark)**
(c) What is the median time? **(1 mark)**
(d) What is the range of the times? **(1 mark)**

Use the actual data values, not just the values of the stems.

Had a look ☐ Nearly there ☐ Nailed it! ☐

PROBABILITY & STATISTICS

Sampling

In statistics, a **population** is a group of people you are interested in. A **sample** is a smaller group chosen from a larger population. You can use data from the sample to make **predictions** about the whole population.

Advantages of sampling

- ✓ It is cheaper to survey a sample than a whole population.
- ✓ It is quicker to collect data from a sample.
- ✓ It is easier to analyse data from a sample and calculate statistics.

Random sample

In a random sample, every member of the population has an equal chance of being included in the sample. Here are two ways of selecting a random sample:

1. Put the names of every member of the population in a hat and select your sample at random.
2. Assign a number to every member of the population and choose random numbers using a computer or calculator.

Problem solved! If a sample is too small the results can be **biased**. This means that the sample is not representative of the whole population, and predictions about the population won't be very accurate. There are two reasons why Ashik's sample might be biased:

Problem	Solution
Ashik only selected a sample of 5 students to represent his entire school.	Choose a **larger** sample. Ashik should have surveyed at least 20 or 30 students.
All the students in Ashik's sample were in one class, so were in the same year group.	Choose a **random sample**. If everyone has an equal chance of being selected there will be a range of ages.

Worked example

Ashik wants to investigate the number of hours of television students in his school watch each week. He surveys five members of his class and gets these results:

15   6   22   11   18

(a) Use this data to estimate the mean number of hours of television watched each week by students in Ashik's school. **(2 marks)**

15 + 6 + 22 + 11 + 18 = 72
72 ÷ 5 = 14.4
An estimate of the mean for the population is 14.4 hours.

(b) Comment on the reliability of your estimate. **(1 mark)**

The estimate is not very reliable because of the small sample size.

(c) How could Ashik reduce bias in his sample? **(2 marks)**

Select a larger sample, and select a random sample from the whole population (school).

Now try this

Amy and Paul are investigating whether trains at their local station run on time. This table summarises their results.

	Amy	Paul
Number of trains observed	6	25
Number of trains late	3	8

Amy says, 'There is a 50:50 chance that a train will be late.'

(a) Comment on the reliability of Amy's statement. **(1 mark)**

(b) Use Paul's results to estimate the probability of a train being late. **(1 mark)**

There is more about probability on page 127.

PROBABILITY & STATISTICS

Had a look ☐ Nearly there ☐ Nailed it! ☐

Comparing data

You can use averages like the **mean** or **median** and measures of spread like the **range** to compare two sets of data. Follow these steps:

1 Calculate the same average and the range for both data sets.

2 Write a sentence for each statistic **comparing** the values for each data set.

3 Only make a statement if you can back it up with **statistical evidence**.

Using graphs and charts

Some graphs and charts are particularly useful for comparing data.

- ✓ A **dual bar chart** lets you compare two frequency distributions.
- ✓ Two **pie charts** can be used to compare **proportions**.
- ✓ A back-to-back **stem and leaf** diagram shows you **all** the data values for two distributions. You can use it to calculate the averages or the range.

Worked example

Target grade 3

Melissa and Fran want to compare their long jump distances. They each jump five times. Fran's distances are:

263 cm   194 cm   220 cm   305 cm   280 cm

Melissa's distances have a mean of 292 cm and a range of 185 cm.

Compare the distances for Melissa and Fran.

(4 marks)

$263 + 194 + 220 + 305 + 280 = 1262$

Fran's mean $= 1262 \div 5 = 252.4$ cm

Fran's range $= 305 - 194 = 111$ cm

Melissa jumped further on average because she had a larger mean.

Fran's jumps were more consistent because she had a smaller range.

You are given the mean and the range for Melissa's distances, so you need to calculate these statistics for Fran's distances.
There is more about calculating the mean and the range on page 120.

Writing conclusions

Here are some examples of good sentences comparing data:

> Class A had more consistent exam results because they had a smaller range.

> Trees in Park B are shorter on average than trees in Park A (smaller median).

> The medians were similar, so on average the apples from both farms were the same weight.

Remember statistics can be the same as well as different.

You need to interpret your results in the context of the question. Write one sentence for the mean and one sentence for the range.

Now try this

Target grade 3

Anna and Carla are in the same maths class. Every week they take a 20-mark test. Here are Anna's first eight test scores.

11   15   9   12   13   16   13   15

Carla's mean score in these tests is 11.

Carla's highest score was 14 and her lowest score was 10.

Compare Anna and Carla's test scores. **(4 marks)**

Start by working out Anna's mean score, and the ranges of both sets of scores.

Had a look ☐   Nearly there ☐   Nailed it! ☐

PROBABILITY & STATISTICS

Hot Topic

Probability 1

The probability that an event will happen is a value from 0 to 1.
The probability tells you how likely the event is to happen.
An event that is **certain** to happen has a probability of 1.
An event that is **impossible** has a probability of 0.
You can write a probability as a fraction, a decimal or a percentage.

Impossible — Even chance — Certain
0 ————————————— 1

Fraction	Decimal	Percentage
$\frac{1}{2}$	0.5	50%

Worked example — Target grade 1

It is **very likely** that it will rain in Newcastle next October. Put a cross near 1 on the probability scale.

(a) On this probability scale, mark with an × the probability that it will rain in Newcastle next October. **(1 mark)**

0 ———————|————×—— 1

(b) Isobel says the probability she will be late for school is 7. Explain why Isobel is wrong. **(1 mark)**

Probabilities are numbers from 0 to 1.

Writing probabilities

The probability of rolling a 6 is $\frac{1}{6}$
You can write P(6) = $\frac{1}{6}$
There is one 6. There are six possible outcomes: 1, 2, 3, 4, 5, 6.

The probability of a coin landing heads up is $\frac{1}{2}$. You can write P(Heads) = $\frac{1}{2}$
There is one head. There are two possible outcomes: heads or tails.

Worked example — Target grade 1

This spinner has eight equal sections.
The spinner is spun.

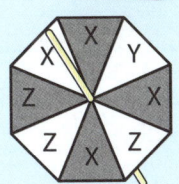

(a) Write down the probability that the spinner will land on the letter Z. **(1 mark)**

$\frac{3}{8}$

(b) Use a suitable probability **word** to complete this sentence:
The chance of the spinner landing on X is _evens_. **(1 mark)**

Golden rule

Probability = $\frac{\text{Number of successful outcomes}}{\text{Total number of possible outcomes}}$

Half the sections on the spinner have the letter X, so there is an even chance of landing on X. Here are some probability words you can use in your exam:

likely unlikely evens certain impossible

Now try this — Target grade 2

Count the number of successful outcomes in each case.

A letter is chosen at random from this word.

| P | H | Y | S | I | C | I | S | T | S |

Work out the probability that the letter will be
(a) I **(1 mark)**   (b) S or Y or C **(1 mark)**   (c) M. **(1 mark)**

Worked solution video

127

PROBABILITY & STATISTICS

Had a look ☐   Nearly there ☐   Nailed it! ☐

Probability 2

The probabilities (P) of all the different outcomes of an event add up to 1.

If you know the probability that something will happen, you can calculate the probability that it won't happen.

P(event doesn't happen) = 1 − P(event does happen)

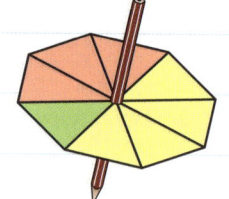

This spinner will definitely land on either red, yellow or green. So the probability of this happening is 1.

The probability of rolling a 6 on a normal fair dice is $\frac{1}{6}$. So the probability of **not** rolling a 6 is $1 - \frac{1}{6} = \frac{5}{6}$

P(Red) + P(Yellow) + P(Green) = 1

Worked example

Target grade 4

A spinner can land on red, blue, white or green. The table shows the probabilities of the spinner landing on each colour.

Colour	Red	Blue	White	Green
Probability	0.12	x	$2x$	0.28

Work out the value of x.   **(3 marks)**

$0.12 + x + 2x + 0.28 = 1$
$0.4 + 3x = 1$ (-0.4)
$3x = 0.6$ $(\div 3)$
$x = 0.2$

Problem solved! The probabilities have to add up to 1. Use this fact to write an equation, then solve it to work out the value of x.

For a reminder about solving linear equations have a look at page 30.

Check it!
P(Blue) = x = 0.2   P(White) = $2x$ = 0.4
$0.12 + 0.2 + 0.4 + 0.28 = 1$ ✓

Expectation

If you flip a coin 100 times, you can expect to get heads about 50 times. You probably won't get heads exactly 50 times, but it's a good guess.

Expected number of outcomes = Number of trials × Probability

You can use expectation to help you decide if a dice or coin is **fair**. These two coins have been flipped 50 times each.

Coin 1

Head	ĦĦ ĦĦ ĦĦ ĦĦ III
Tail	ĦĦ ĦĦ ĦĦ ĦĦ ĦĦ II

About the same number of heads and tails. This coin is probably **fair**.

Coin 2

Head	ĦĦ ĦĦ IIII
Tail	ĦĦ ĦĦ ĦĦ ĦĦ ĦĦ ĦĦ ĦĦ I

A lot more than the expected number of tails. This coin is probably **biased**.

Now try this

Worked solution video

Preti has a packet of sweets. The sweets are red, yellow, green or orange.

Preti picks one sweet at random. This table gives the probability that the sweet will be red, yellow or orange.

Target grade 1

Colour	Red	Yellow	Green	Orange
Probability	0.13	0.36		0.28

Complete the table by working out the probability that the sweet will be green.

(2 marks)

Had a look ☐ Nearly there ☐ Nailed it! ☐

PROBABILITY & STATISTICS

Relative frequency

You need to be able to calculate probabilities for data given in graphs and tables. You can use this formula to estimate a probability from a frequency table:

Probability = $\dfrac{\text{Frequency of outcome}}{\text{Total frequency}}$

When a probability is calculated like this it is sometimes called a **relative frequency**.

Golden rule
Probability estimates based on relative frequency are **more accurate** for larger samples (or for more trials in an experiment).

Worked example

Target grade 4

An egg farm weighed a sample of 40 eggs. It recorded the results in a frequency table.

Weight, w (g)	Frequency
$45 \leq w < 50$	6
$50 \leq w < 55$	9
$55 \leq w < 60$	15
$60 \leq w < 65$	10

(a) Roselle buys some eggs from the farm and picks one at random. Estimate the probability that the egg weighs 55 g or more. **(2 marks)**

$P(w \geq 55) \approx \dfrac{25}{40}$

(b) Comment on the accuracy of your estimate. **(1 mark)**

40 is a fairly small sample size, so the estimate is not very accurate.

Examiners' report

In the sample there were $15 + 10 = 25$ eggs which weighed 55 g or more, out of a total of 40 eggs. Don't round any values here – your answer is an **estimate** because it is based on a sample.

Real students have struggled with questions like this in recent exams – **be prepared!**

Experimental probability

You can carry out an experiment to estimate the probability of something happening. This table shows the results of throwing a drawing pin 60 times.

Number of trials	10	20	30	40	50	60
Frequency of landing point up	8	11	17	25	30	37

To estimate the probability that the drawing pin will land point up, you calculate the relative frequency. The most accurate estimate will be based on the largest number of trials.

Now try this

Target grade 4

A four-sided dice is rolled 40 times. The results are shown in the table.

Number	1	2	3	4
Frequency	8	4	21	7

(a) Work out the estimated probability of getting a 3. **(1 mark)**

(b) Work out the theoretical probability of getting a 3 on a fair four-sided dice. **(1 mark)**

(c) Do you think this dice is fair? Give a reason for your answer. **(1 mark)**

Worked solution video

If a dice is fair, the experimental probability gets closer to the theoretical probability as the number of trials increases.

PROBABILITY & STATISTICS

Had a look ☐   Nearly there ☐   Nailed it! ☐

Frequency and outcomes

You might need to record all the possible **outcomes** of two or more events. You can use a **frequency tree** to show the frequencies of each possible outcome.

Worked example   Target grade 3

Wilfred owns 80 books. He has **not** read 15 of these books.
20 of the books he **has** read are hardbacks.
He has 52 paperbacks in total. The rest of the books are hardbacks.

(a) Use the information to complete this frequency tree.

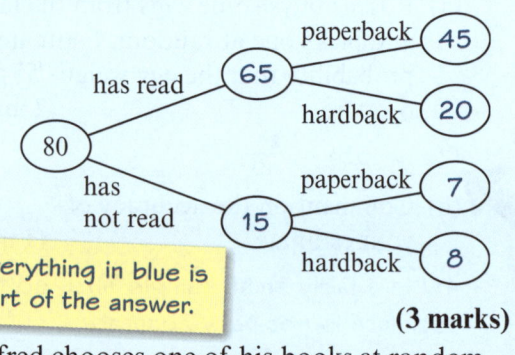

Everything in blue is part of the answer.

(3 marks)

Wilfred chooses one of his books at random.
(b) Work out the probability that it is a paperback he has not read.   (2 marks)

$\frac{7}{80}$

Golden rule

In a frequency tree, each frequency is equal to the sum of its branches.

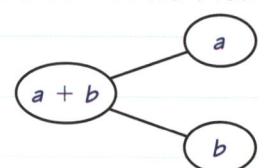

1. Write 15 in the bottom left oval, then use the golden rule above. Wilfred has read $80 - 15 = 65$ books.
2. He has $80 - 52 = 28$ hardbacks in total. 20 of these are books he **has** read, so 8 must be books he has **not** read.
3. Use the golden rule to complete the frequency tree.

Sample space diagrams

A **sample space diagram** shows you all the possible outcomes of two events. Here are all the possible outcomes when two coins are flipped.

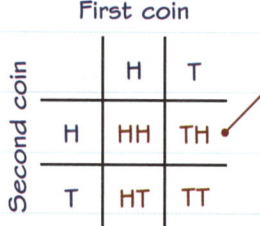

There are four possible outcomes. TH means getting a tail on the first coin and a head on the second coin.

Worked example   Target grade 1

This bag contains 30 counters.
They are all either black or white.
A counter is chosen at random.
The probability that it is black is $\frac{1}{5}$. How many white counters are in the bag?   (2 marks)

$P(\text{White}) = 1 - \frac{1}{5} = \frac{4}{5}$
$\frac{4}{5} \times 30 = 24$

Check it!
There are $30 - 24 = 6$ black counters in the bag.
$P(\text{Black}) = \frac{6}{30}$ ✓

Now try this   Target grade 3

Dawn is eating a meal at a restaurant. She is in a rush so she chooses one main course and one dessert from the menu at random.
(a) Write down all the possible outcomes.   (2 marks)
(b) Work out the probability that Dawn chooses fish pie and cheesecake.   (3 marks)

Main courses
Steak and chips
Fish pie
Mushroom pasta

Desserts
Cheesecake
Lemon mousse
Tiramisu

Had a look ☐ Nearly there ☐ Nailed it! ☐

PROBABILITY & STATISTICS

Venn diagrams

You can use a Venn diagram to show frequencies in a probability question. This Venn diagram shows the results when 50 people were asked whether they owned a dog (D) or a cat (C). The rectangle represents everyone who was surveyed. The **number** in each section tells you **how many** people that section represents.

In total 21 + 8 + 6 + 15 = 50 people were surveyed. This symbol represents all of them.

This oval represents people who owned a cat.

Some people owned a cat and a dog so the ovals overlap.

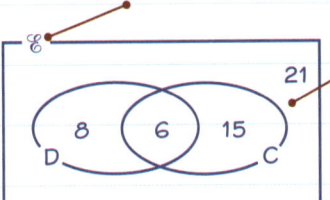

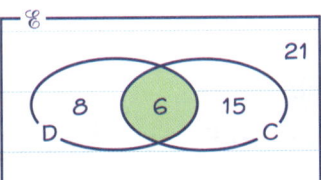

6 people owned a dog **and** a cat. You can write this as D ∩ C.
∩ means **and** or **intersection**.

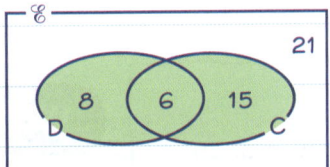

8 + 6 + 15 = 29 people owned a dog **or** a cat. You can write this as D ∪ C.
∪ means **or** or **union**.

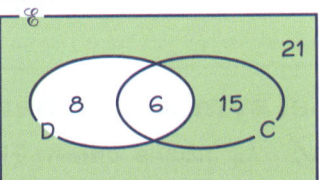

15 + 21 = 36 people **did not** own a dog. You can write this as D'. D' means **not** D or the **complement** of D.

Worked example

Target grade 5

36 members of a youth club were surveyed about the sports they played.
19 members played tennis.
14 members played football.
6 members played both tennis and football.

(a) Draw a Venn diagram to show this information. **(3 marks)**

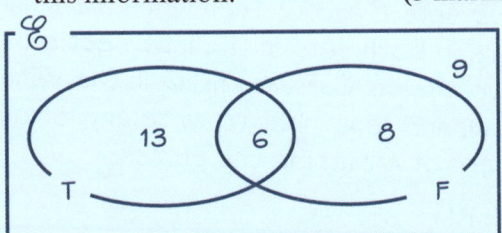

(b) One of the members is chosen at random. Write down the probability that the member plays neither tennis nor football. **(1 mark)**

$\frac{9}{36} = \frac{1}{4}$

T ∩ F
Fill in the centre of the Venn diagram first. 6 members play both sports.

T ∩ F'
The 19 members who play tennis include the 6 members who play both sports.
So 19 − 6 = 13 members who play tennis but *not* football.

T' ∩ F
14 − 6 = 8 members who play football but *not* tennis.

T' ∩ F'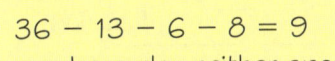
36 − 13 − 6 − 8 = 9 members play neither sport.

Check it!
13 + 6 = 19 members play tennis ✓
8 + 6 = 14 members play football ✓
13 + 6 + 8 + 9 = 36 members surveyed ✓

Now try this

Target grade 5

The Venn diagram shows information about how 30 people in an office travel to work.
One person is chosen at random. Find the probability that this person
(a) uses both the bus and the train on their journey **(1 mark)**
(b) uses the bus at some point in their journey. **(2 marks)**

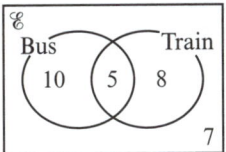

PROBABILITY & STATISTICS

Had a look ☐ Nearly there ☐ Nailed it! ☐

Set notation

In mathematics a **set** is a collection of **members** (or **elements**). The elements in a set could be numbers, words or letters. You can define a set in two different ways:

1 Listing the elements
   A = {onions, carrots, peas} — You use curly brackets to define a set.
   B = {13, 14, 15, 16} — Members are separated by commas.

2 Using a rule
   C = {months with exactly 30 days} — 'June' is a member of this set.
   D = {odd numbers between 10 and 20} — You could also write set D as {11, 13, 15, 17, 19}.

Set symbols to learn

✓ ∪ means **union**. The union of two sets is the set of elements that belong to **either** set.

✓ ∩ means **intersection**. The intersection of two sets is the set of elements that belong to **both** sets.

✓ ℰ means the **universal** set. It represents all the elements you are allowed to consider in a question.

✓ A' means **not A** or the **complement** of A. It is everything in ℰ but not in A.

Worked example — Target grade 5

X = {multiples of 3 between 7 and 19}
Y = {even numbers between 7 and 19}

(a) List the members of X ∪ Y. (2 marks)

8, 9, 10, 12, 14, 15, 16, 18

Karl says that 15 is a member of X ∩ Y.

(b) Is Karl correct? You must give a reason for your answer. (2 marks)

No. The members of X ∩ Y must be in both sets. 15 is in X but it is an odd number so it is not in Y.

Worked example — Target grade 5

ℰ = {1, 2, 3, 4, 5, 6, 7, 8, 9, 10, 11, 12}
P = {multiples of 3}
Q = {4, 5, 6, 7, 10, 12}

(a) Complete the Venn diagram for this information. (4 marks)

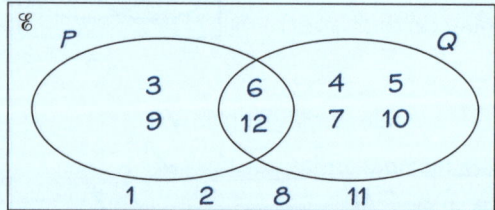

A number is chosen at random from the universal set ℰ.

(b) What is the probability that it is in the set P ∩ Q. (2 marks)

$\frac{2}{12}$

Start by labelling the two circles to represent the two sets P and Q. Then write the numbers that are in **both** sets in the **intersection** of the two circles. Carry on filling in the Venn diagram, and remember to write any leftover members of ℰ outside the circles.

Check it!
You should have written every member of ℰ exactly once each.

Now try this — Target grade 5

ℰ = {1, 2, 3, 4, 5, 6, 7, 8, 9, 10, 11, 12}
A = {factors of 20}
B = {prime numbers less than 20}

(a) List the members of the set A ∩ B. (2 marks)

(b) Is 7 a member of A ∪ B? Explain your answer. (2 marks)

(c) Show all of this information on a Venn diagram. (4 marks)

Had a look ☐   Nearly there ☐   Nailed it! ☐

PROBABILITY & STATISTICS

Tricky Topic

Independent events

Two events are **independent** if the outcome of one does not affect the outcome of the other. You can work out the probability of two independent events **both** occurring by **multiplying** the probabilities.

Worked example *Target grade 5*

Huan spins this spinner. He keeps spinning until he lands on green. Work out the probability that Huan spins the spinner

(a) exactly twice **(2 marks)**

P(blue) × P(green) = $\frac{7}{8} \times \frac{1}{8} = \frac{7}{64}$

(b) more than twice. **(2 marks)**

P(blue) × P(blue) = $\frac{7}{8} \times \frac{7}{8} = \frac{49}{64}$

Problem solved! (a) Huan stops when he lands on green, so in order to spin the spinner **exactly** twice he must land on blue on the first spin, **and** green on the second spin.

(b) In order for Huan to spin more than twice he must land on blue on **both** of the first two spins.

Tree diagrams

You can show independent events on a tree diagram.

You write the probabilities on the branches. For each pair of branches, the probabilities add up to 1:
$\frac{1}{8} + \frac{7}{8} = 1$

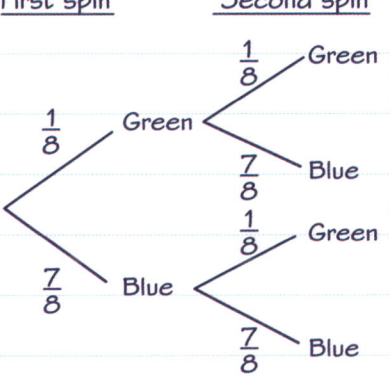

Work out probabilities using these rules:

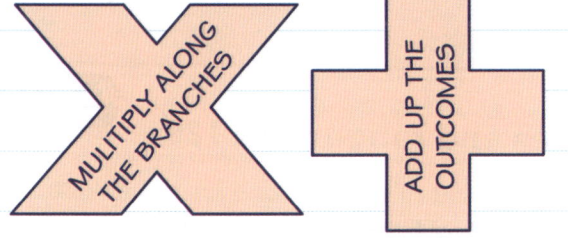

MULTIPLY ALONG THE BRANCHES   ADD UP THE OUTCOMES

Worked example *Target grade 5*

Aidan and Chloe each take a basketball shot. The probability that Aidan scores is 0.3 The probability that Chloe scores is 0.4 Work out the probability that exactly one player scores. **(3 marks)**

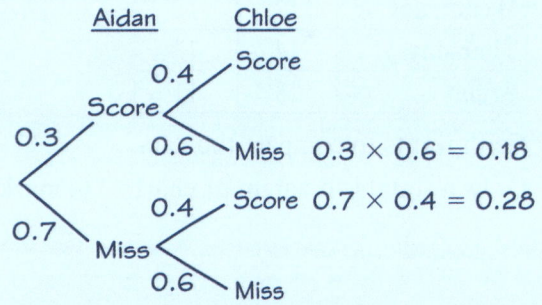

0.3 × 0.6 = 0.18
0.7 × 0.4 = 0.28

P(Exactly one player scores)
= 0.3 × 0.6 + 0.7 × 0.4
= 0.18 + 0.28
= 0.46

There are **two** successful outcomes. Multiply along the branches to work out the probability of each one, then add up the answers.

Now try this *Target grade 5*

Ravina and Anjali are going to school one morning.
The probability that Ravina will arrive late is 0.2.
The probability that Anjali will arrive late is 0.5.

(a) Draw a probability tree diagram to show this information. **(3 marks)**

(b) Work out the probability that Ravina and Anjali will both arrive late for school. **(2 marks)**

PROBABILITY & STATISTICS

Had a look ☐   Nearly there ☐   Nailed it! ☐

Problem-solving practice 1

About half of the questions in your Foundation GCSE exam will require you to **problem-solve, reason, interpret** or **communicate** mathematically. If you come across a tricky or unfamiliar question in your exam you can try some of these strategies.

- ✓ Sketch a diagram to see what is going on.
- ✓ Try the problem with smaller or easier numbers.
- ✓ Plan your strategy before you start.
- ✓ Write down any formulae you might be able to use.
- ✓ Use x or n to represent an unknown value.

AO2

AO3

Now try this

1 The table shows information about the numbers of Year 7 pupils absent from Keith's school last week.

	Boys	Girls
Monday	8	10
Tuesday	11	9
Wednesday	12	12
Thursday	14	13
Friday	13	11

Keith wants to compare the data.
Draw a suitable diagram or chart. **(4 marks)**

Bar charts page 117
Line graphs page 123

In this question **you** have to choose what type of diagram or chart to use. It is best to use a bar chart or a line graph.

TOP TIP

- Label **both** axes correctly.
- Draw a key for boys and girls, or make sure it is clear which bars (or lines) are for boys and which are for girls.

---

2 Abi has five cards.
Each card has a number written on it.

The mean of the five numbers is 6.
One of the numbers is hidden.
Work out the hidden number. **(2 marks)**

Averages and range page 120

You could try some different values for the hidden number and work out the mean each time. But you can save time by using the rule in the Top tip. There are five numbers and the mean is 6, so the sum of the numbers must be $5 \times 6 = 30$.

TOP TIP

$$\text{Mean} \times \text{Number of data values} = \text{Sum of data values}$$

Problem-solving practice 2

Now try this

3 Some students in a class weighed themselves. Here are their results.
Boys' weights in kg
70 65 45 52 63 72 63
Girls' weights in kg
65 45 47 61 44 67 55 56 63
Compare fully the weights of these students. **(6 marks)**

Averages and range page 120 — Target grade 3

There are **6 marks** for this question.
To give a full answer you need to **compare** the data. So (1) calculate averages like the mean or median and a measure of spread like the range, and (2) write a sentence for each of these, comparing the boys and the girls.

TOP TIP

You need to **calculate** first, then use your calculations to write your **conclusion**. It's safest to write in full sentences for your conclusion.

---

4 The scatter graph shows the French and German marks of 15 students.

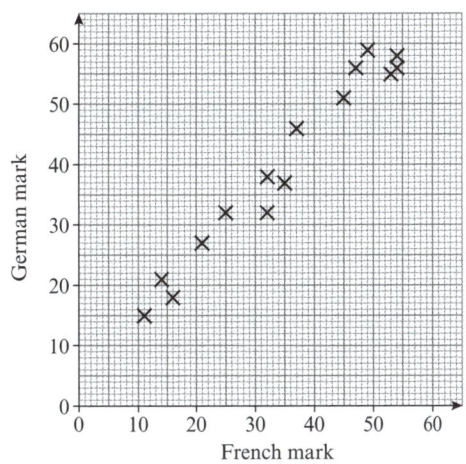

Jade's French mark was 42.
Estimate Jade's German mark. **(2 marks)**

Scatter graphs page 119 — Target grade 4

Start by drawing a line of best fit on the scatter graph. Next draw a line up from 42 on the horizontal axis to your line of best fit. Then draw a line across to the vertical axis to estimate Jade's German mark. Make sure you draw the lines you use on your graph to show your working.

TOP TIP

When you are reading information from a graph you should always give your answer to the nearest small square.

---

5 Amy picks one card at random from group A and one card at random from group B.

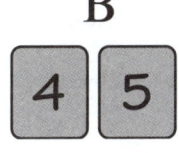

She adds together the numbers on the two cards. What is the probability that her total is

(a) 8   (b) 7   (c) an odd number?

(4 marks)

Independent events page 133
Counting strategies page 19 — Target grade 5

You can tackle this in two different ways.
1. Write down all the possible outcomes when two cards are picked.
2. Consider the two picks as two separate independent events, and multiply the probabilities of each.

TOP TIP

In a probability question make sure you have considered all the possible outcomes, and all the successful outcomes.

Answers

NUMBER

1. Place value
1 £974, £1497, £1749, £1947, £1974
2 (a)

Car	A	B	C	D	E
bought	£2200	£7800	£4200	**£10 700**	£6500
sold	£2900	£9000	**£2700**	£11 200	£5750
profit or loss	£700 profit	**£1200 profit**	£1500 loss	£500 profit	**£750 loss**

(b) £150 profit

2. Negative numbers
1 (a) −6 (b) 5
2 (a) 48 (b) −6
3 −3 and −5, −10 and 2, −9 and 1

3. Rounding numbers
1 (a) 3800 (b) 3760
2 (a) 50 (b) 3000 (c) 0.009 (d) 7
3 0.180
4 The cooker could be more than 76 cm wide (e.g. 76.3 cm).

4. Adding and subtracting
1 (a) 1714 (b) 1028
2 (a) 376 (b) 172
3 £1.57

5. Multiplying and dividing
1 (a) 7200 (b) 2304 (c) 2262
2 (a) 156 (b) 129
3 $18 \times 6 = 108$, $7 \times 15 = 105$. Yes, she has enough coins.

6. Decimals and place value
1 (a) 8 tenths (b) 8 thousandths
2 0.508, 0.51, 0.517, 0.571, 0.58
3 (a) 8.736 (b) 130

7. Operations on decimals
1 (a) 20.43 (b) 11.912
2 (a) 6.78 (b) 195.84
3 (a) £239.40 (c) 94p

8. Squares, cubes and roots
1 (a) 36 (b) 5 (c) 64 (d) 11
2 e.g. $25 = 5^2$, $9 = 3^2$, $25 \times 9 = 225$ which is not even

9. Indices
1 (a) 7^{10} (b) 7^7 (c) 7^3 (d) 7^{15}
2 (a) 5^5 (b) 64
3 (a) 1 (b) $\frac{1}{4}$ (c) $\frac{1}{16}$ (d) $\frac{4}{25}$ (e) $\frac{64}{27} = 2\frac{10}{27}$

10. Estimation
1 $\frac{80 \times 300}{60 \times 40} = 10$
2 (a) 108 (or 81, rounding $\frac{4}{3}$ to 1)
  (b) $3 < 3.14$ and $3 < 3.2$ (and $1 < \frac{4}{3}$) so underestimate

11. Factors, multiples and primes
1 (a) 21 (b) 15 (c) 24 (d) 2, 37
2 $280 = 2^3 \times 5 \times 7$

12. HCF and LCM
1 (a) $132 = 2^2 \times 3 \times 11$, $110 = 2 \times 5 \times 11$
  (b) 22 (c) 660
2 (a) $3^2 \times 5$ (b) $2 \times 3^3 \times 5^2 \times 7^4$

13. Fractions
1 £8 profit
2 e.g. $10 \times 0.25 + 20 \times 0.3 = 8.5$, so Sandeep makes a 50p profit

14. Operations on fractions
1 (a) $\frac{8}{9}$ (b) $\frac{9}{16}$
2 (a) $\frac{3}{8}$ (b) $\frac{7}{55}$
3 (a) $\frac{1}{6}$ (b) $\frac{14}{15}$

15. Mixed numbers
1 (a) $5\frac{5}{6}$ (b) $2\frac{6}{25}$
2 7

16. Calculator and number skills
1 (a) 4 (b) 7 (c) 17 (d) 125
2 (a) 31 (b) 1.4

17. Standard form 1
1 (a) 2.8×10^{-4} (b) 391 000
2 (a) 2.71×10^6 (b) 1.75×10^7

18. Standard form 2
1 (a) 2.7×10^{12} (b) 3.2×10^{-6}

19. Counting strategies
1 AY, AZ, BY, BZ, CY, CZ
2 15

20. Problem-solving practice 1
1 10p
2 e.g.
  Cost of theatre trip
  $£15 \times 25 + £11.50 \times 5 + £5.75 \times 20 = £547.50$
  Cost of zoo trip
  $£240 + £18 \times 5 + £12 \times 20 = £570$
  Conclusion
  The lowest possible total cost is the theatre with circle tickets: £547.50

21. Problem-solving practice 2
3 $\frac{7}{20}$
4 $2 \times \frac{3}{8} = \frac{3}{4}$ of a bag per day.
  $14 \div \frac{3}{4} = 18\frac{2}{3}$. Susan can feed the dogs for 18 days from 1 bag.
5 96 mm

ALGEBRA

22. Collecting like terms
1. (a) $6x$ (b) $8q$ (c) $10t$ (d) $3n$
2. term
3. $9p - 3q$

23. Simplifying expressions
(a) n^5 (b) $6r^3$ (c) $32yz$ (d) $4f$ (e) 7

24. Algebraic indices
1. (a) m^9 (b) k^7
2. (a) h^8 (b) t^{20} (c) a^5
3. (a) y^8 (b) $125p^6$

25. Substitution
1. 12
2. 6
3. (a) 13 (b) 51 (c) 23

26. Formulae
1. 36 cm^2
2. 49

27. Writing formulae
1. (a) $50x + 35y$ (b) $T = 50x + 35y$
2. (a) $A = 3s^2$ (b) 75 cm^2

28. Expanding brackets
1. (a) $3y - 18$ (b) $m^2 + 7m$
2. (a) $13a - 6b$ (b) $-w + 21$ (c) $m^2 + 22m$

29. Factorising
1. (a) $6(n + 3)$ (b) $3(p - 1)$
2. (a) $4t(t + 2)$ (b) $3x(x - 4)$
3. $2, 2p, p - 2q$

30. Linear equations 1
1. (a) $y = 8$ (b) $m = 17$ (c) $k = 19$
2. (a) $n = 5$ (b) $w = 10.5$
3. (a) $n = 36$ (b) $q = 77$

31. Linear equations 2
1. (a) $x = 5$ (b) $x = 0.5$
2. (a) $t = -11$ (b) $h = -0.5$
3. (a) $y = 14.5$ (b) $m = -1$

32. Inequalities
1. (a) $x > -3$ (b) $x \leq 4$
2. (a)

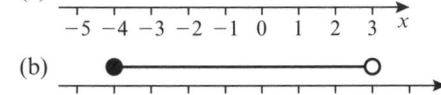

(b)

33. Solving inequalities
1. (a) $x \geq 5$ (b) $4 > x$ (or $x < 4$)
2. (a) $-2, -1, 0, 1, 2$ (b) $0, 1, 2, 3$

34. Sequences 1
1. 33, 45
2. (a) 51
   (b) 72 (11th term)
3. 21, 34, 55

35. Sequences 2
(a) $4n - 1$
(b) No. 22nd term = 87, 23rd term = 91
   (or $4n - 1 = 89$
   $n = 22.5$ which is not a whole number)

36. Coordinates
(a) (3, 6)
(b) 6 units
(c) (6, 4)

37. Gradients of lines
(a) 5
(b) No. The gradient is 5 so for every £10 000 spent revenue increases by approximately £50 000.

38. Straight-line graphs 1
$y = -2x + 2$

39. Straight-line graphs 2
1. $y = 6x - 20$
2. $y = 2x - 1$

40. Real-life graphs
(a) £5
(b) £15
(c) Gradient = 2.5 so additional cost per minute = 2.5p. So additional 10 minutes cost 25p. Claim is true.

41. Distance–time graphs
(a) 30 minutes
(b) 30 km/h

42. Rates of change
A: False. The car was accelerating between 0 and 5 seconds.
B: True: The graph is flat between 5 and 15 seconds.
C: True: The graph slopes upwards between 0 and 5 seconds.
D: False: The car was travelling at a constant speed between 5 and 15 seconds.

43. Expanding double brackets
1. $(x + 6)(x + 3) = x^2 + 3x + 6x + 18$
   $= x^2 + 9x + 18$
2. (a) $x^2 + 4x - 5$
   (b) $p^2 - 12p + 36$

44. Quadratic graphs
(a)

x	−3	−2	−1	0	1	2	3
y	11	6	3	2	3	6	11

(b)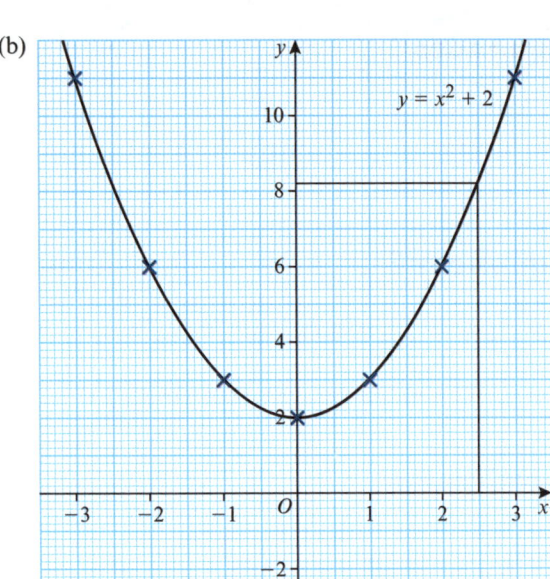

(c) $y = 8.2$ (to the nearest small square)

45. Using quadratic graphs
(a) $x = -0.2$ and $x = 4.2$
(b) $x = -1.6$ and $x = 5.6$

46. Factorising quadratics
1 (a) $(x + 2)(x + 4)$ (b) $(x - 2)(x - 8)$
2 (a) $(x - 12)(x + 12)$ (b) $(x + 7)(x - 7)$

47. Quadratic equations
1 (a) $(x + 3)(x + 2)$ (b) $x = -2, x = -3$
2 (a) $x = 0, x = 5$
   (b) $x = 4, x = -7$
   (c) $x = 12, x = -12$

48. Cubic and reciprocal graphs
1 (a) D   (b) E   (c) A   (d) B   (e) F   (f) C

49. Simultaneous equations
1 $x = 5, y = 1.5$
2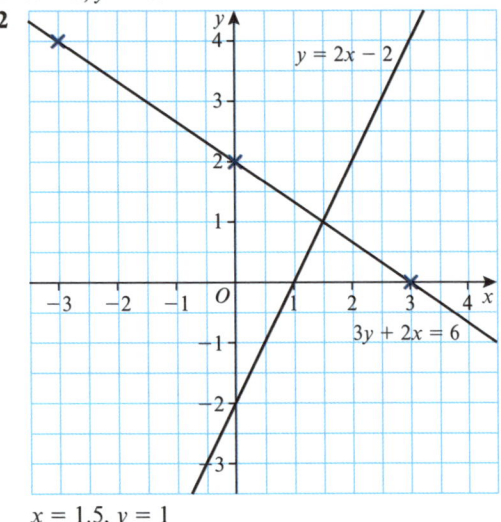
   $x = 1.5, y = 1$

50. Rearranging formulae
1 (a) $Q = \dfrac{W + 100}{7}$ (b) $V = \dfrac{m}{d}$
2 $n = \dfrac{2m - p}{2}$

51. Using algebra
1 $(x - 3) + 2x + (x - 3) + 2x = x + (x + 4) + x + (x + 4)$
   $6x - 6 = 4x + 8$
   $2x = 14$
   $x = 7$
   Rectangle A: width = 4 m, height = 14 m
2 e.g. $n = 4$: $n^2 + 4n + 3 = 35 = 5 \times 7$

52. Identities and proof
1 $(a + b)(a - b) = a^2 - ab + ab - b^2$
   $= a^2 - b^2$
2 $4x - 4n = 3x - 1$
   $x = 4n - 1$
   $4n$ is an even number so $4n - 1$ is an odd number (so x is an odd number)

53. Problem-solving practice 1
1 e.g. $A(10, 10), B(50, 10), C(50, 30), D(10, 30)$
2 1.4 kg

54. Problem-solving practice 2
3 $t = 3.5$
4 nth term $= 6n + 3$
   $6 \times 10 + 3 = 63$
   $6 \times 11 + 3 = 69$
   So 65 is not a term in the sequence.
5 3 minutes $= \dfrac{3}{60} = \dfrac{1}{20}$ hours
   Speed $= \dfrac{\text{distance}}{\text{time}} = y \div \dfrac{1}{20} = y \times \dfrac{20}{1} = 20y$
6 (a) e.g. when $x = 10$: $y = 8 \times 10 - 30 = 50$, so $(10, 50)$ is a point on the line.
   $y = 8x - 30$
   (b) e.g. the answer is reliable because it uses algebra

RATIO & PROPORTION

55. Percentages
1 £546
2 35%

56. Fractions, decimals and percentages
1 (a) $\dfrac{3}{20}$ (b) $\dfrac{17}{25}$
2 36%, $\dfrac{2}{5}$, 0.42
3 $5 \times 12 = 60$ counters in total
   35% of 60 = 21
   $\dfrac{2}{5}$ of 60 = 24
   $60 - 21 - 24 = 15$
   Haydon wins 15 counters

57. Percentage change 1
1 1351
2 34.6% (1 d.p.)

58. Percentage change 2
Cruks Cameras price = £171.50
Spivs Cameras price = £175.50
Cruks Cameras is cheaper.

59. Ratio 1
1 (a) $\frac{4}{7}$ (b) $2:3$
2 $7:10$
3 56 counters

60. Ratio 2
1 £120
2 Rice : onion $= 300 : 160 = 15 : 8$

61. Metric units
1 (a) 320 cm (b) 250 ml (c) 96 cm (d) 1.7 kg
2 1.7 mm

62. Reverse percentages
1 £45
2 £220 000

63. Growth and decay
(a) £5304.50 (b) 4 years

64. Speed
1 4 hours and 15 minutes
2 266 miles

65. Density
9.234 g/cm^3

66. Other compound measures
1 16.6 km/l
2 2.4 minutes or 2 minutes and 24 seconds

67. Proportion
1 (a) £2.80 (b) £6.72
2 16 days

68. Proportion and graphs
(a) 4 inches (b) 32 cm
(c) Graph is a straight line and passes through the origin $(0, 0)$

69. Problem-solving practice 1
1 45
2 Sportcentre: £45   Footwear First: £60   Action Sport: £48
Sportcentre is the cheapest.

70. Problem-solving practice 2
3 312
4 £3610.80
5 24.72 litres

GEOMETRY & MEASURES

71. Symmetry
(a) e.g.  (b) e.g.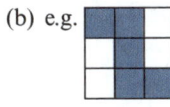

72. Quadrilaterals
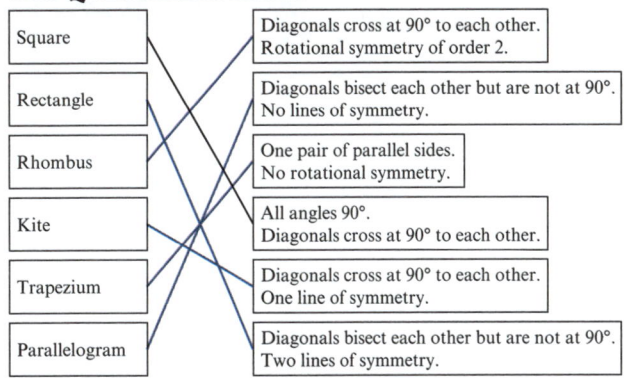

73. Angles 1
$b = 55°$ (Vertically opposite angles are equal).
$c = 125°$ (Angles on a straight line add up to 180°).
$d = 125°$ (Vertically opposite angles are equal).

74. Angles 2
(a) $x = 134°$ (Corresponding angles are equal).
(b) $y = 97°$

75. Solving angle problems
1 (a) 71° (b) 24°
2 $56 + y + (y + 60) = 180$
$\quad\quad 2y + 116 = 180$
$\quad\quad\quad\quad 2y = 64$
$\quad\quad\quad\quad\quad y = 32$
Angle $y = 32°$

76. Angles in polygons
144°

77. Time and timetables
(a) 14:36 (b) 8 hours

78. Reading scales
25 g

79. Perimeter and area
34 cm^2

80. Area formulae
$x = 8$

81. Solving area problems
(a) 3 cm (b) 5 cm (c) 58 cm^2 (d) 50 cm

82. 3-D shapes
(a) 250 cm^2
(b) $50 \times 250 = 12\,500$
$12\,500 \div 3000 = 4\frac{1}{6}$
Karl must buy 5 cans.

83. Volumes of cuboids
6.5 cm

84. Prisms
(a) 1200 cm^3 (b) 920 cm^2

85. Units of area and volume
1. (a) $23\,000\,\text{cm}^2$ (b) $0.4\,\text{cm}^3$
2. $3.5 \times 10^8\,\text{mm}^3$
3. $37\,500\,\text{N/m}^2$

86. Translations
(a)
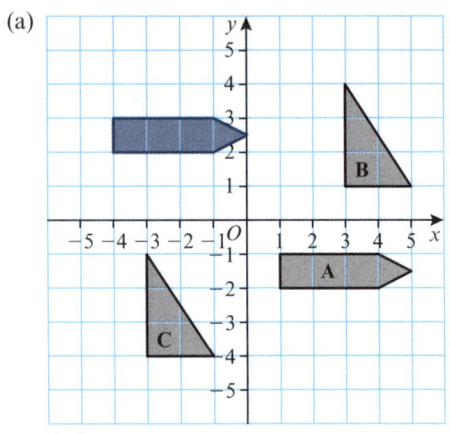
(b) $\begin{pmatrix} -6 \\ -5 \end{pmatrix}$

87. Reflections
(a), (b), (c)
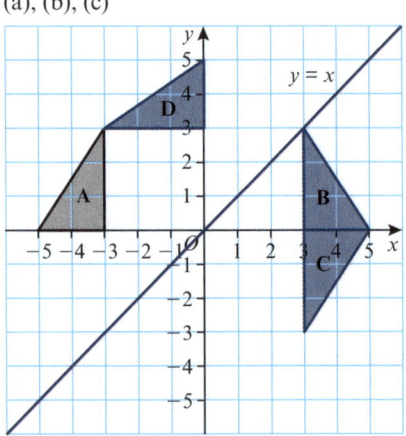
(d) Reflection in the line $y = -x$

88. Rotations
(a) Rotation 90° anticlockwise about (4, 3)
(b) Rotation 180° about (6, 3)

89. Enlargements
(a), (b)
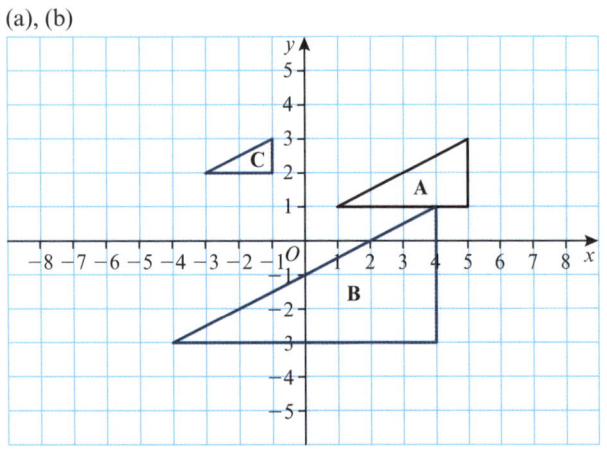

90. Pythagoras' theorem
(a) $y = 3.5\,\text{cm}$ (b) $z = 9.1\,\text{cm}$

91. Line segments
1. 9.22 units (2 d.p.)
2. 18.56 units (2 d.p.)

92. Trigonometry 1
(a) 44.8° (b) 56.2° (c) 48.6°

93. Trigonometry 2
(a) 4.4 cm (b) 5.4 cm (c) 15.7 cm

94. Exact trigonometry values
$\cos 60° = \dfrac{2.4}{AC}$

$AC \times \cos 60° = 2.4$

$AC = \dfrac{2.4}{\cos 60°}$

$= 2.4 \div \tfrac{1}{2} = 4.8\,\text{m}$

95. Measuring and drawing angles
1. (a) 28° (b) 308°
2.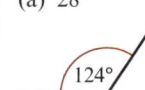

96. Measuring lines
1. (a) 2.8 cm (b) 5.5 cm (c) 4.9 cm
   (d) 6.0 cm
2. Accept between 20 and 25 feet

97. Plans and elevations
(a) (i)  (ii)  (iii)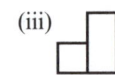

(b) $30\,\text{cm}^2$

98. Scale drawings and maps
80 cm

99. Constructions 1
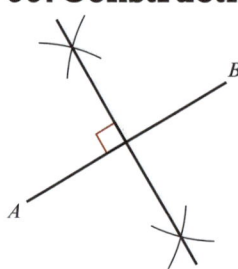

100. Constructions 2
1
2
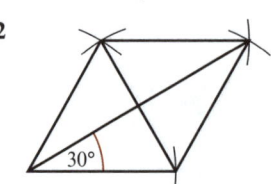

101. Loci

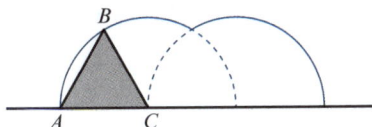

102. Bearings
(a) 280 km (accept 275 km to 290 km)
(b) 225°

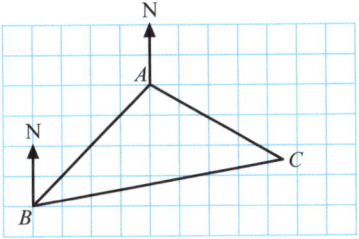

103. Circles
(a) 38.2 cm
(b) 19.1 cm

104. Area of a circle
9.4 cm

105. Sectors of circles
12.6 cm (3 s.f.)

106. Cylinders
55 litres

107. Volumes of 3-D shapes
Volume of cone = $\frac{1}{3}\pi r^2 h = 3\pi h$
Volume of sphere = $\frac{4}{3}\pi r^3 = 36\pi$
$3\pi h = 36\pi$ ($\div 3\pi$)
$h = 12$ cm

108. Surface area
1 52π cm^2
2 115 cm^2 (3 s.f.)

109. Similarity and congruence
A and K, H and J, D and G

110. Similar shapes
(a) 130°   (b) 31.5 cm

111. Congruent triangles
Angle ABC = Angle ADC = 90°
Hypotenuse of both triangles is a common side (AC) so they are equal
$BC = CD$
So RHS is satisfied and triangles are congruent.

112. Vectors
(a) $\begin{pmatrix} 6 \\ 10 \end{pmatrix}$ (b) $\begin{pmatrix} 5 \\ -4 \end{pmatrix}$ (c) $\begin{pmatrix} -7 \\ -24 \end{pmatrix}$

113. Problem-solving practice 1
1 $10y + 90° = 360°$
    $y = 27°$
2 13.8 km (1 d.p.)

114. Problem-solving practice 2
3 e.g. Radius of large semicircle = $2x$
    Radius of small semicircle = x
    Shaded area = $\frac{1}{2}\pi x^2 + \frac{1}{2}\pi x^2 = \pi x^2$
    Area of large semicircle = $\frac{1}{2}\pi(2x)^2 = 2\pi x^2$
    which is twice the shaded area
4 5.8 m (1 d.p.)
5 Interior angle of A = 360° − 90° − 108° = 162°
    Exterior angle of A = 18°
    $n = 360° \div 18° = 20$

PROBABILITY & STATISTICS

115. Two-way tables

	Gloss	Matt	Lustre	Total
Small	20	35	12	**67**
Medium	63	**102**	29	**194**
Large	**22**	24	**18**	64
Total	105	**161**	59	325

116. Pictograms
(a) Key: 🏠 = £50 000

Lancashire	🏠🏠🏠
Herefordshire	🏠🏠🏠🏠🏠
Essex	🏠🏠🏠🏠🏠
Surrey	🏠🏠🏠🏠🏠🏠🏠🏠
Lincolnshire	🏠🏠🏠

(b) £250 000

117. Bar charts
56 applications

118. Pie charts
Salad 84°, Stew 36°, Pizza 180°, Curry 60°

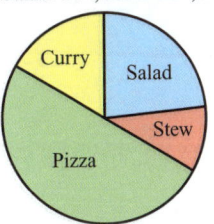

119. Scatter graphs
(a) Accept 23.5°C to 24.5°C
(b) This is not a reliable estimate because it is extrapolation.

120. Averages and range
(a) 14   (b) 7   (c) 13   (d) 12

121. Averages from tables 1
(a) 1 goal   (b) 2 goals   (c) $95 \div 50 = 1.9$ goals

122. Averages from tables 2
(a) $10 < x \leq 20$ (b) $20 < x \leq 30$ (c) $1350 \div 50 = 27$

123. Line graphs
(a)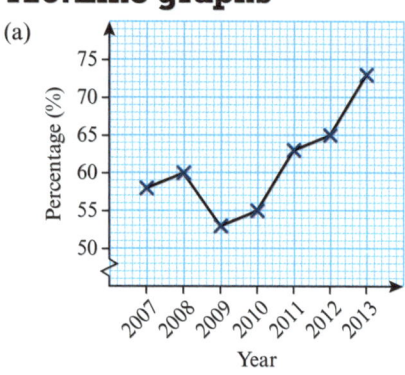

(b) Upward trend (or upward trend from 2009 onwards)

124. Stem-and-leaf diagrams
(a)
```
0 | 8 9 9
1 | 0 1 2 2 5 6 8 9     Key 1|2 represents 12 minutes
2 | 3 4 4 4 6 8
```
(b) 24 minutes   (b) 16 minutes   (c) 20 minutes

125. Sampling
(a) Amy only observed 6 trains so her statement is unreliable.
(b) $\frac{8}{25} = 0.32$

126. Comparing data
Anna's mean $104 \div 8 = 13$ Anna's range $16 - 9 = 7$
Carla's mean 11   Carla's range $14 - 10 = 4$
On average, Anna scored better than Carla because she had a higher mean.
Carla's scores were more consistent than Anna's because she had a smaller range.

127. Probability 1
(a) $\frac{2}{10}$ or $\frac{1}{5}$ (b) $\frac{5}{10}$ or $\frac{1}{2}$ (c) 0

128. Probability 2

Colour	Red	Yellow	Green	Orange
Probability	0.13	0.36	**0.23**	0.28

129. Relative frequency
(a) $\frac{21}{40}$ (b) $\frac{1}{4}$
(c) No. $\frac{21}{40}$ is more than twice $\frac{1}{4} = \frac{10}{40}$ so the dice is not fair.

130. Frequency and outcomes
(a) SC, SL, ST, FC, FL, FT, MC, ML, MT
(b) $\frac{1}{9}$

131. Venn diagrams
(a) $\frac{5}{30} = \frac{1}{6}$ (b) $\frac{15}{30} = \frac{1}{2}$

132. Set notation
(a) 2, 5
(b) Yes, because it is a prime number less than 20 (so it is in either A or B).

(c)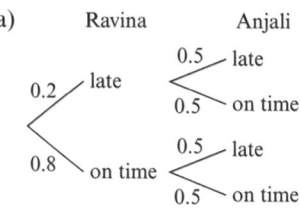

133. Independent events
(a)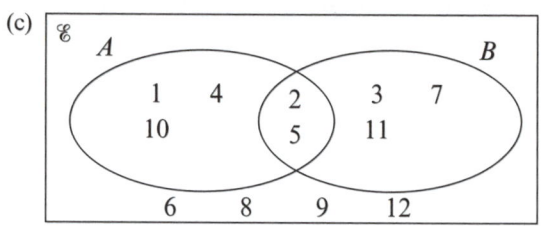

(b) P(Both late) $= 0.2 \times 0.5 = 0.1$

134. Problem-solving practice 1
1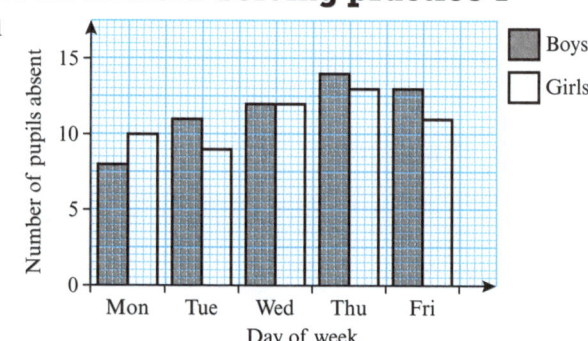

2 9

135. Problem-solving practice 2
3 e.g.

	Mean	Median	Range
Boys	61.4 kg	63 kg	27 kg
Girls	55.9 kg	56 kg	23 kg

The boys had a higher mean and median than the girls, so they were on average heavier. The girls had a smaller range which indicates that their weights were more consistent than the boys' weights.

4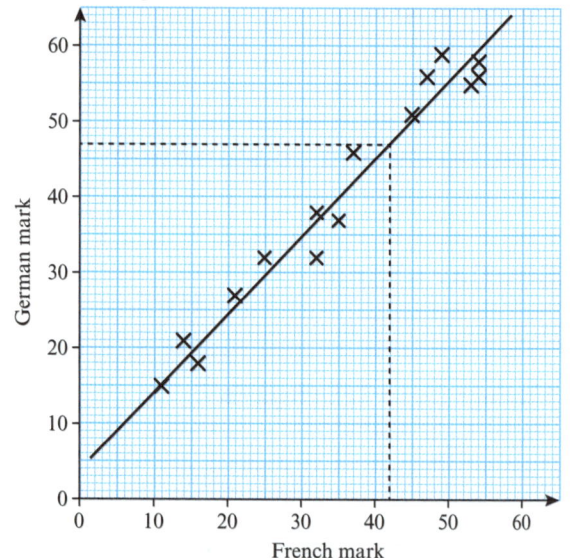

Estimated German mark is 47

5 (a) $\frac{1}{6}$ (b) $\frac{2}{6} = \frac{1}{3}$ (c) $\frac{3}{6} = \frac{1}{2}$

Notes

Notes

Notes

Notes

Notes

Published by Pearson Education Limited, 80 Strand, London, WC2R 0RL.

www.pearsonschoolsandfecolleges.co.uk

Copies of official specifications for all Pearson qualifications may be found on the website: qualifications.pearson.com

Text © Harry Smith and Pearson Education Limited 2015
Original illustrations © Pearson Education Limited 2015
Typeset and illustrations by Newgen KnowledgeWorks (P) Ltd, Chennai, India and Tech-Set Ltd, Gateshead
Produced by Out of House Publishing
Cover illustration by Kamae Design Ltd

The right of Harry Smith to be identified as author of this work has been asserted by him in accordance with the Copyright, Designs and Patents Act 1988.

First published 2015

23

18

British Library Cataloguing in Publication Data
A catalogue record for this book is available from the British Library

ISBN 978 1 447 98804 5

Copyright notice
All rights reserved. No part of this publication may be reproduced in any form or by any means (including photocopying or storing it in any medium by electronic means and whether or not transiently or incidentally to some other use of this publication) without the written permission of the copyright owner, except in accordance with the provisions of the Copyright, Designs and Patents Act 1988 or under the terms of a licence issued by the Copyright Licensing Agency, 5th Floor, Shackleton House, Hay's Galleria, 4 Battle Bridge Lane, London, SE1 2HX (www.cla.co.uk). Applications for the copyright owner's written permission should be addressed to the publisher.

Printed by Bell and Bain Ltd, Glasgow

Notes from the publisher
1. In order to ensure that this resource offers high-quality support for the associated Pearson qualification, it has been through a review process by the awarding body. This process confirms that this resource fully covers the teaching and learning content of the specification or part of a specification at which it is aimed. It also confirms that it demonstrates an appropriate balance between the development of subject skills, knowledge and understanding, in addition to preparation for assessment.

Endorsement does not cover any guidance on assessment activities or processes (e.g. practice questions or advice on how to answer assessment questions), included in the resource nor does it prescribe any particular approach to the teaching or delivery of a related course.

While the publishers have made every attempt to ensure that advice on the qualification and its assessment is accurate, the official specification and associated assessment guidance materials are the only authoritative source of information and should always be referred to for definitive guidance.

Pearson examiners have not contributed to any sections in this resource relevant to examination papers for which they have responsibility.

Examiners will not use endorsed resources as a source of material for any assessment set by Pearson.

Endorsement of a resource does not mean that the resource is required to achieve this Pearson qualification, nor does it mean that it is the only suitable material available to support the qualification, and any resource lists produced by the awarding body shall include this and other appropriate resources.

2. Pearson has robust editorial processes, including answer and fact checks, to ensure the accuracy of the content in this publication, and every effort is made to ensure this publication is free of errors. We are, however, only human, and occasionally errors do occur. Pearson is not liable for any misunderstandings that arise as a result of errors in this publication, but it is our priority to ensure that the content is accurate. If you spot an error, please do contact us at resourcescorrections@pearson.com so we can make sure it is corrected.